Unmanned Aircraft Design

A Review of Fundamentals

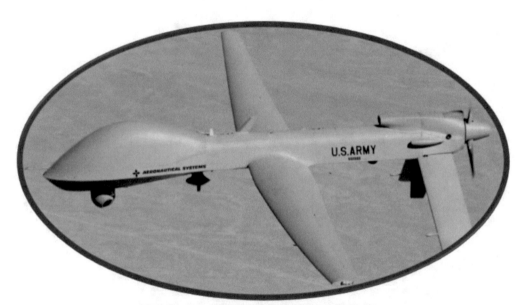

ER/MP Gray Eagle: Enhanced MQ-1C Predator

Synthesis Lectures on Mechanical Engineering

Synthesis Lectures on Mechanical Engineering series publishes 60–150 page publications pertaining to this diverse discipline of mechanical engineering. The series presents Lectures written for an audience of researchers, industry engineers, undergraduate and graduate students. Additional Synthesis series will be developed covering key areas within mechanical engineering.

Introduction to Refrigeration and Air Conditioning Systems: Theory and Applications
Allan Kirkpatrick
September 2017

Resistance Spot Welding: Fundamentals and Applications for the Automotive Industry
Menachem Kimchi and David H. Phillips
September 2017

Unmanned Aircraft Design: Review of Fundamentals
Mohammad Sadraey
September 2017

MEMS Barometers Towards Vertical Position Detection: Background Theory, System Prototyping, and Measurement Analysis
Dimosthenis E. Bolankis
2017

Vehicle Suspension System Technology and Design
Amir Khajepour and Avesta Goodarzi
April 2017

Engineering Finite Element Analysis
Ramana Pidaparti
May 2017

Unmanned Aircraft Design: A Review of Fundamentals
Mohammad Sadraey

ISBN: 978-3-031-79581-7   print
ISBN: 978-3-031-79582-4   ebook

DOI 10.1007/978-3-031-79582-4

A Publication in the Springer series
SYNTHESIS LECTURES ON MECHANICAL ENGINEERING, #04

Series ISSN: 2573-3168 Print   2573-3176 Electronic

Unmanned Aircraft Design

A Review of Fundamentals

Mohammad H. Sadraey
Southern New Hampshire University

SYNTHESIS LECTURES ON MECHANICAL ENGINEERING #04

ABSTRACT

This book provides fundamental principles, design procedures, and design tools for unmanned aerial vehicles (UAVs) with three sections focusing on vehicle design, autopilot design, and ground system design. The design of manned aircraft and the design of UAVs have some similarities and some differences. They include the design process, constraints (e.g., g-load, pressurization), and UAV main components (autopilot, ground station, communication, sensors, and payload). A UAV designer must be aware of the latest UAV developments; current technologies; know lessons learned from past failures; and they should appreciate the breadth of UAV design options.

The contribution of unmanned aircraft continues to expand every day and over 20 countries are developing and employing UAVs for both military and scientific purposes. A UAV system is much more than a reusable air vehicle or vehicles. UAVs are air vehicles, they fly like airplanes and operate in an airplane environment. They are designed like air vehicles; they have to meet flight critical air vehicle requirements. A designer needs to know how to integrate complex, multi-disciplinary systems, and to understand the environment, the requirements and the design challenges and this book is an excellent overview of the fundamentals from an engineering perspective.

This book is meant to meet the needs of newcomers into the world of UAVs. The materials are intended to provide enough information in each area and illustrate how they all play together to support the design of a complete UAV. Therefore, this book can be used both as a reference for engineers entering the field or as a supplementary text for a UAV design course to provide system-level context for each specialized topic.

KEYWORDS

unmanned aerial vehicles, design, automatic flight control system, autopilot, drone, remotely piloted vehicle

Contents

Preface

The Unmanned Aerial Vehicle (UAV) is a remotely piloted or self-piloted aircraft that can carry cameras, sensors, communications equipment or other payloads. All flight operations (including take-off and landing) are performed without on-board human pilot. In some reports of DOD, Unmanned Aircraft System (UAS) is preferred. In media reports, Drone is preferred. Mission is to perform critical missions without risk to personnel and more cost effectively than comparable manned system.

The contributions of unmanned aircraft in sorties, hours, and expanded roles continue to increase. As of September 2004, some 20 types of coalition UAVs, large and small, have flown over 100,000 total flight hours in support of Operation Enduring Freedom and Operation Iraqi Freedom. Their once reconnaissance only role is now shared with strike, force protection, and signals collection. These diverse systems range in cost from a few hundred dollars (Amazon sells varieties) to tens of millions of dollars. Range in capability from Micro Air Vehicles (MAV) weighing much less than a pound to aircraft weighing over 40,000 pounds.

The UAV system includes four elements: (1) air vehicle; (2) ground control station; (3) payload; and (4) maintenance/support system. The design of manned aircraft and the design of UAVs have some similarities; and some differences. They include the: (1) design process; (2) constraints (e.g., g-load, pressurization; and (3) UAV main components (autopilot, ground station, communication system, sensors, and payload). A UAV designer must be aware of: (a) latest UAV developments; (b) current technologies; (c) known lessons learned from past failures; and (d) designer should appreciate the breadth of UAV design options.

A design process requires both integration and iteration. A design process includes: (1) Synthesis: the creative process of putting known things together into new and more useful combinations; (2) Analysis: the process of predicting the performance or behavior of a design candidate; and (3) Evaluation: the process of performance calculation and comparing the predicted performance of each feasible design candidate to determine the deficiencies. A designer needs to know how to integrate complex, multi-disciplinary systems, and to understand the environment, the requirements and the design challenges.

The objectives of this book are to review the design fundamentals of Unmanned Aerial Vehicles. It will have three Parts and ten Chapters. Part I (Chapters 1 and 2) is on "Vehicle Design" and covers design fundamentals, and design disciplines. This part covers UAV classifications, design project planning, decision making, feasibility analysis, systems engineering approach, design groups, design phases, design reviews, evaluation, feedback, aerodynamic design, structural design, propulsion system design, landing gear design, mechanical systems design, and control surfaces design.

Part II (Chapters 3–7) is dedicated to the Autopilot Design. It will cover dynamic modeling, control system design, navigation system design, guidance system design, and microcontroller. This part will discuss the topics such as: aircraft aerodynamic forces and moments, stability and control derivatives, transfer function model, state-space model, aircraft dynamics, linearization, fundamentals of control systems, control laws, conventional design techniques, optimal control, robust control, digital control, stability augmentation, coordinate systems, inertial navigation, way-point navigation, sensors, avionics, gyroscopes, GPS, navigation laws, guidance laws, proportional navigation guidance, line-of-sight guidance, lead angle, tracking a command, flight path stabilization, turn coordination, command systems, modules/components, flight software, integration, and full autonomy. A few advanced topics such as detect (i.e., sense)-and-avoid, automated recovery, fault monitoring, intelligent flight planning, and manned-unmanned teaming will also be reviewed in this part.

In Part III (Chapters 8, 9, and 10), equipment design is presented which includes ground control station communication systems, payloads, and launch and recovery. The following topics will be discussed: ground element types, portable ground station, mission control elements, remote control personnel, support equipment, transportation, coordination, hardware and software, radio frequencies, elements of communication system, communication techniques, transmitters, receivers, telemetry, measurement devices, antennas, radar, civil payloads, military payloads, disposable payloads, imagery equipment, payload handling, payload management, payload-structure integration, conventional launch, rail launchers, hand launch, air launch, and recovery systems. Due to the limited length of this book, many topics are reviewed in brief.

Putting a book together requires the talents of many people, and talented individuals abound at Morgan & Claypool Publishers. My sincere gratitude goes to Paul Petralia, Executive Editor of Engineering, and Deb Gabriel for composition. My special thanks go to the outstanding copy editor and proof-reader who are essential in creating an error-free text. I especially owe a large debt of gratitude to my students and the reviewers of this text. Their questions, suggestions, and criticisms have helped me to write more clearly and accurately and have markedly influenced the evolution of this book.

Mohammad Sadraey
July 2017

[Unattributed figures are held in the public domain and are from either the U.S. Government Departments or Wikipedia.]

Part I

CHAPTER   1

Design Fundamentals

1.1   INTRODUCTION

The Unmanned Aerial Vehicle (UAV) is a remotely piloted or self-piloted aircraft that can carry payloads such as cameras, sensors, and communications equipment. All flight operations (including take-off and landing) are performed without on-board human pilot. In some reports of DOD, Unmanned UAV System (UAS) is preferred. In media reports, the term "drone" is utilized. The UAV mission is to perform critical flight operations without risk to personnel and more cost effectively than comparable manned system. A civilian UAV is designed to perform a particular mission at a lower cost or impact than a manned aircraft equivalent.

UAV design is essentially a branch of engineering design. Design is primarily an analytical process which is usually accompanied by drawing/drafting. Design contains its own body of knowledge that is independent of the science-based analysis tools that is usually coupled with it. Design is a more advanced version of a problem solving technique that many people use routinely.

Research in unmanned aerial vehicles (UAVs) has grown in interest over the past couple decades. There has been tremendous emphasis in unmanned aerial vehicles, both of fixed and rotary wing types over the past decades. Historically, UAVs were designed to maximize endurance and range, but demands for UAV designs have changed in recent years. Applications span both civilian and military domains, the latter being the more important at this stage. Early statements about performance, operation cost, and manufacturability are highly desirable already early during the design process. Individual technical requirements have been satisfied in various prototype, demonstrator and initial production programs like Predator, Global Hawk, and other international programs. The possible break-through of UAV technology requires support from the aforementioned awareness of general UAV design requirements and their consequences on cost, operation and performance of UAV systems.

In June of 2016, the Department of Transportation's Federal Aviation Administration has finalized the first operational rules for routine commercial use of small unmanned aircraft systems [27], opening pathways toward fully integrating UAS into the nation's airspace. These new regulations aim to harness new innovations safely, to spur job growth, advance critical scientific research and save lives. Moreover, in June of 2017, European Commission has released a blueprint for UAV standards which will "unify laws across the EU" by creating a common low-level airspace called the U-space that covers altitudes of up to 150 m.

The design principles for UAVs are similar to the principles developed over the years and used successfully for the design of manned UAV. The size of UAV varies according to the purpose of their utility. In many cases the design and constructions of UAVs faces new challenges and, as a result of these new requirements, several recent works are concerned with the design of innovative UAVs. Autonomous vehicle technologies for small and large fixed-wing UAVs are being developed by various startups and established corporations such as Lockheed Martin. A number of conceptual design techniques, preliminary design methodologies, and optimization has been applied to the design of various UAVs including Medium Altitude Long Endurance (MALE) UAV using multi-objective genetic algorithm.

The first UAV designs that appeared in the early nineties were based on the general design principles for full UAV and findings of experimental investigations. The main limitation of civil UAV's is often low cost. An important area of UAV technology is the design of autonomous systems. The tremendous increase of computing power in the last two decades and developments of general purpose reliable software packages made possible the use of full configuration design software packages for the design, evaluation, and optimization of modern UAV.

UAVs are air vehicles, they fly like airplanes and operate in an airplane environment. They are designed like air vehicles. They have to meet flight critical air vehicle requirements. You need to know how to integrate complex, multi-disciplinary systems. You need to understand the environment, the requirements and the design challenges.

A UAV system is much more than a reusable air vehicle or vehicles. The UAV system includes five basic elements: (1) the Environment in which the UAV(s) or the Systems Element operates (e.g., the airspace, the data links, relay UAV, etc.); (2) the air vehicle(s) or the Air Vehicle Element; (3) the control station(s) or the Mission Control Element; (4) the payload(s) or the Payload Element; and (5) the maintenance and support system or the Support Element.

The design of manned UAV and the design of UAVs have some similarities; and some differences such as: design process; constraints (e.g., g-load, pressurization); and UAV main components (autopilot, ground station, communication system, sensors, payload). A UAV designer must be aware of the: (1) latest UAV developments; (2) current technologies; and (3) known lessons learned from past failures. Designers should appreciate breadth of UAV design options.

UAV are not new, they have a long history in aviation. Their history stretches back to the First World War (1920s), Cold War, Korean War, Vietnam War (RPV), Yugoslavia, Afghanistan, First and Second Persian Gulf war, and other wars (e.g., Pakistan, Yemen, Syria, and Africa). At least 20 countries are using or developing over 76 different types of UAVs. The contributions of unmanned UAV in sorties, hours, and expanded roles continue to increase. As of September 2004, some 20 types of coalition UAVs, large and small, have flown over 100,000 total flight hours in support of Operation Enduring Freedom and Operation Iraqi Freedom. Their once reconnaissance-only role is now shared with strike, force protection, and signals collection.

In this chapter, definitions, design process, UAV classifications, current UAVs, and challenges will be covered. In addition, conceptual design, preliminary design, and detail design of a UAV based on systems engineering approach are introduced. In each stage, application of this approach is described by presenting the design flow chart and practical steps of design.

1.2 UAV CLASSIFICATIONS

It is a must for a UAV designer to be aware of classifications of UAVs which is based on various parameters such as cost, size, weight, mission, and the user. For instance, UAV ranges in weight from Micro Air Vehicles (MAV) weighing less than 1 pound to UAV weighing over 40,000 lb. Moreover, these diverse systems range in cost from a few hundred dollars (Amazon sells varieties) to tens of millions of dollars (e.g., Global Hawk). In addition, UAV missions ranges from reconnaissance, combat, target acquisition, electronic warfare, surveillance, special purpose UAV, target and decoy, relay, logistics, research and development, and civil and commercial UAVs, to environmental application (e.g., University of Kansas North Pole UAV for measuring ice thickness).

The early classification includes target drones and remotely piloted vehicles (RPVs). The current classification ranges from Micro UAVs (less than 15 cm long, or 1 lb); to High-altitude Long Endurance (HALE); to tactical and combat UAVs. In this section, characteristics of various classifications are briefly presented.

The Micro Unmanned Aerial Vehicles (MAV) was originally a DARPA program to explore the military relevance of Micro Air Vehicles for future military operations, and to develop and demonstrate flight enabling technologies for very small UAV (less than 15 cm/6 in. in any dimension). The Tactical UAV (e.g., Outrider) is designed to support tactical commanders with near-real-time imagery intelligence at ranges up to 200 km. The Joint Tactical UAV (Hunter) was developed to provide ground and maritime forces with near-real-time imagery intelligence at ranges up to 200 km. The Medium Altitude Endurance UAV (Predator) provides imagery intelligence to satisfy Joint Task Force and Theater Commanders at ranges out to 500 nautical miles. The High Altitude Endurance UAV (Global Hawk) is intended for missions requiring long-range deployment and wide-area surveillance or long sensor dwell over the target area. Table 1.1 shows the UAV classifications from a few aspects including size, mass, and mission. The MLB Bat 4, a mini-UAV (Figure 2.7) with a length of 2.4 m, a wingspan of 3.9 m, and a maximum takeoff mass of 45 kg has a maximum cruising speed [54] of 120 knot.

In the U.S. military, the classification is mainly based on a tier system. For instance, in the U.S. Air Force the Tier I is for low altitude, long endurance missions, while Tier II is for medium altitude; long endurance (MALE) missions (e.g., Predator). Moreover, Tier II+ is for high-altitude, long-endurance (HALE) missions and Tier III- denotes HALE low observable. For other military forces, the following is the classification. Marine Corp: Tier I: Mini UAV; (e.g., Wasp, and MLB

Bat); Tier II: (e.g., Pioneer); and Tier III: Medium range, (e.g., Shadow). Army: Tier I: Small UAV, (e.g., Raven); Tier II: Short range, tactical UAV, (e.g., Shadow 200); and Tier III: Medium range, tactical UAV.

Table 1.1: **UAV classification**

No.	Class	Mass	Size	Normal Operating Altitude	Range	Endurance
1	Micro	< 0.2 lb	< 10 cm	< 50 ft	0.1-0.5 km	< 1 hr
2	Mini	0.2-1 lb	10-30 cm	< 100 ft	0.5-1 km	< 1 hr
3	Very small	2-5 lb	30-50 cm	< 1000 ft	1-5 km	1-3 hr
4	Small	5-20 lb	0.5-2 m	1,000-5,000 ft	10-100 km	0.5-2 hr
5	Medium	100-1,000 lb	5-10 m	10,000-15,000 ft	500-2,000 km	3-10 hr
6	Large	10,000-30,000 lb	20-50 m	20,000-40,000 ft	1,000-5,000 km	10-20 hr
7	Tactical/combat	1,000-20,000 lb	10-30 m	10,000-30,000 ft	500-2,000 km	5-12 hr
8	MALE	1,000-10,000 lb	15-40 m	15,000-30,000 ft	20,000-40,000 km	20-40 hr
9	HALE	> 5,000 lb	20-50 m	50,000-70,000 ft	20,000-40,000 km	30-50 hr

Another basis for UAVs classifications in military is echelon: Class 1 supports platoon echelon, (e.g., Raven), micro air vehicle (MAV), and small UAV; Class 2 supports company echelon, (e.g., Interim Class 1 and 2 UAV); Class 3 supports battalion echelon, (e.g., Shadow 200 Tactical UAV); and Class 4 supports unit of action (brigade), (e.g., Hunter), Extended Range/Multipurpose (ER/MP) UAV.

Some current U.S. UAVs [46] are listed here: (1) Army UAV Systems: RQ-1L I-GNAT Organization; RQ-5/MQ-5 Hunter Aerial Reconnaissance Company; RQ-7 Shadow Aerial Reconnaissance Platoon; RQ-11 Raven Team. (2) Air Force UAV Systems: RQ-4 Global Hawk; RQ/MQ-1 Predator; MQ-9 Predator B; Force Protection Aerial Surveillance System, Desert Hawk (Figure 8.3). (3) Navy UAV Systems: RQ-2 Pioneer; RQ-8B Fire Scout. (4) Marine Corps UAV Systems: FQM-151 Pointer; Dragon Eye; Silver Fox; Scan Eagle. (5) Coast Guard UAV Systems: Eagle Eye. (6) Special Operations Command UAV Systems: CQ-10 SnowGoose; FQM-151 Pointer; RQ-11 Raven; Dragon Eye.

It will be very helpful to know the features of some old and current UAVs. **Hunter (RQ-5):** Range: 125 km; Max speed: 110 knots; Dimensions: length: 22.6 ft; span: 29.2 ft; Endurance: 10 hr; Weights: Max Takeoff: 1600 lb; Ceiling: 16,000 ft. Hunter, was cancelled in January 1996 after some 20 air vehicle crashes. **Pioneer RQ-2A:** First flight: 1985; Dimensions: length: 14 ft span: 16.9 ft; Max. TO Weight: 450 lb; Speeds: cruise: 65 knots; dash: 110 knots; it was used extensively in Falujeh, Iraq, 2006. During Operations Desert Shield, the U.S. deployed 43 Pioneers that flew 330 sorties, completing over 1,000 flight hours. In 10 years, Pioneer system has flown nearly 14,000 flight hours. Since 1994, it has flown over Bosnia, Haiti, and Somalia.

Outrider: First flight: 2,000; Range: 200 km; Wing span: 11.1 ft; MTOW: 385 lb; Ceiling: 15,000 ft; Max speed: 110 knot; Endurance: 7.2 hr. **Predator RQ-1A** (Figure 5.4)**:** First flight: 1994; Endurance: 25 hr; Ceiling: 26,000; Payload: 450 lb; Cruise Speed: 90 knots; MTOW: 2100 lb; Wing span: 48.4 ft. Extensively employed in Iraq, Afghanistan, Pakistan, ... **Predator RQ-1B** (Figure 4.10)**:** Honeywell TPE-331-10T, flat-rated to 750 shp; 4,500 kg take-off gross weight; Max speed/altitude: 210 knot/50Kft; - 20 m wingspan; Triplex systems; 1,360 kg fuel; 340 kg internal payload; 1360 kg external payload; 6 store stations/14 Hellfire missiles.

Figure 1.1: Northrop Grumman RQ-4 Global Hawk.

- **Global Hawk RQ-4** (Figure 1.1)**:** First flight was in 1998; Endurance: 41 hr; Ceiling: 65,000; Payload: 2,000 lb; Ranges: 14,000 nm; Cruise Speed: 345 knots; MTOW: 25,500 lb; Wing span: 116 ft. The Defense Advanced Research Projects Agency (DARPA) developed Global Hawk to provide military field commanders with a high-altitude, long-endurance system that can obtain high-resolution, near-real-time imagery of large geographic areas. Flew for the first time at Edwards Air Force Base, California, on Saturday, February 28, 1998. The entire mission, including the take-off and landing, was performed autonomously by the UAV based on its mission plan. The launch and recovery element of the system's ground segment continuously monitored the status of the flight.

1.3   DESIGN PROJECT PLANNING

In order for a design project schedule to be effective, it is necessary to have some procedures for monitoring progress; and in a broader sense for encouraging personnel to progress. An effective general form of project management control device is the Gantt chart is. It presents a project overview which is almost immediately understandable to non-systems personnel; hence it has great value as a means of informing management of project status. A Gantt chart has three main features.

1. It informs the manager and chief designer of what tasks are assigned and who has been assigned to them.

2. It indicated the estimated dates on which tasks are assumed to start and end, and it represents graphically the estimated ration of the task.

3. It indicates the actual dates on which tasks were started and completed and pictures this information.

Like many other planning/management tools, Gantt charts provide the manager/chief designer with an early warning if some jobs will not be completed on schedule and/or if others are ahead of schedule. Gantt charts are also helpful in that they present graphically immediate feedback regarding estimates of personnel skill and job complexity. A Gantt chart provides the chief designer with a scheduling method and enables him/her to rapidly track and assess the design activities on a weekly/monthly basis. An aircraft project such as Global Hawk (Figure 1.1) will not be successful without a design project planning.

1.4   DECISION MAKING

Not every design parameters is the outcome of a mathematical/technical calculations. There are UAV parameters which are determined through a selection process. In such cases, the designer should be aware of the decision making procedures. The main challenge in decision making is that there are usually multiple criteria along with a risk associated with each one. Any engineering selection must be supported by logical and scientific reasoning and analysis. The main challenge in decision making is that there are usually multiple criteria along with a risk associated with each one. There are no straightforward governing equations to be solved mathematically.

A designer must recognize the importance of making the best decision and the adverse of consequence of making the poorest decision. In majority of the design cases, the best decision is the right decision, and the poorest decision is the wrong one. The right decision implies the design success, and the wrong decision results in a fail in the design. As the level of design problem complexity and sophistication increases in a particular situation, a more sophisticated approach is needed.

1.5   DESIGN CRITERIA, OBJECTIVES, AND PRIORITIES

One of the preliminary tasks in UAV configuration design is identifying system design considerations. The definition of a need at the system level is the starting point for determining customer requirements and developing design criteria. The requirements for the system as an entity are established by describing the functions that must be performed. Design criteria constitute a set of "design-to" requirements, which can be expressed in both qualitative and quantitative terms. Design criteria are customer specified or negotiated target values for technical performance measures. These requirements represent the bounds within which the designer must "operate" when engaged in the iterative process of synthesis, analysis, and evaluation. Both operational functions (i.e., those required to accomplish a specific mission scenario, or series of missions) and maintenance and support functions (i.e., those required to ensure that the UAV is operational when required) must be described at the top level.

Various UAV designer have different priorities in their design processes. These priorities are based on different objectives, requirements ,and mission. There are primarily three groups of UAV designers, namely: (1) military UAV designers, (2) civil UAV designers, and (3) homebuilt UAV designers. These three groups of designers have different interests, priorities, and design criteria. There are ten main figures of merit for every UAV configuration designer. They are: (1) production cost, (2) UAV performance, (3) flying qualities, (4) design period, (5) beauty (for civil UAV) or scariness (for military UAV), (6) maintainability, (7) producibility, (8) UAV weight, (9) disposability, and (10) stealth requirement. Table 1.2 demonstrates objectives and priorities of each UAV designer against some figures of merit.

In design evaluation, an early step that fully recognizes design criteria is to establish a baseline against which a given alternative or design configuration may be evaluated. This baseline is determined through the iterative process of requirements analysis (i.e., identification of needs, analysis of feasibility, definition of UAV operational requirements, selection of a maintenance concept, and planning for phase-out and disposal). The mission that the UAV must perform to satisfy a specific customer should be described, along with expectations for cycle time, frequency, speed, cost, effectiveness, and other relevant factors. Functional requirements must be met by incorporating design characteristics within the UAV and its configuration components.

No	Objective	Basis for measurement	Criterion	Units
Table 1.2: Design objectives				
1	Inexpensive in market	Unit manufacturing cost	Manufacturing cost	Dollar
2	Inexpensive in operation	Fuel consumption per km	Operating cost	Liter/km
3	Light weight	Total weight	Weight	N
4	Small size	Geometry	Dimensions	m
5	Fast	Speed of operation	Performance	km/hr
6	Maintainable	Man-hour to maintain	Maintainability	Man-hour
7	Producible	Required technology for man-ufacturing	Manufacturability	-
8	Recyclable	Amount of hazardous or non-recyclable materials	Disposability	kg
9	Maneuverable	Turn radius; turn rate	Maneuverability	m
10	Detect and avoid	Navigation sensors	Guidance and control	
11	Airworthiness	Safety standards	Safety	-
12	Autonomy	Autopilot complexity	Crashworthiness/for-mation flight	-

As an example, Table 1.3 illustrates three scenarios of priorities (in percent) for military UAV designers. Among ten figures of merit (or criteria), grade "1" is the highest priority and grade "10" is the lowest priority. The grade "0" in this table means that, this figure of merit is not a criterion for this designer. The number one priority for a military UAV designer is UAV performance, while for a homebuilt UAV designer cost is the number one priority. It is also interesting that stealth capability is an important priority for a military UAV designer, while for other three groups of designers, it is not important at all. These priorities (later called weights) reflect the relative importance of the individual figure of merit in the mind of the designer.

Design criteria may be established for each level in the system hierarchical structure. The optimization objectives must be formulated in order to determine the optimum design. A selected UAV configuration would be optimum based on only one optimization function. Applicable criteria regarding the UAV should be expressed in terms of technical performance measures and should be prioritized at the UAV (system) level. Technical performance measures are measures for character-istics that are, or derive from, attributes inherent in the design itself. It is essential that the develop-ment of design criteria be based on an appropriate set of design considerations, considerations that lead to the identification of both design-dependent and design-independent parameters, and that support the derivation of technical performance measures.

No	Figure of Merit	Priority	Designer # 1	Designer # 2	Designer # 3
	Table 1.3: Three scenarios of priorities (in percent) for a military UAV designer				
1	Cost	4	8	9	9
2	Performance	1	50	40	30
3	Autonomy	2	10	15	20
4	Period of design	5	7	7	8
5	Scariness	10	1	1	2
6	Maintainability	7	4	5	5
7	Producibility	6	6	6	7
8	Weight	8	3	4	4
9	Disposability	9	2	2	3
10	Stealth	3	9	11	12
		Total	100%	100%	100%

1.6   FEASIBILITY ANALYSIS

In the early stages of design and by employing brainstorming, a few promising concepts are suggested which seems consistent with the scheduling and available resources. Prior to committing resources and personnel to the detail design phase, an important design activity—feasibility analysis—must be performed. There are a number of phases through which the system design and development process must invariably pass. Foremost among them is the identification of the customer-related need and, from that need, the determination of what the system is to do. This is followed by a feasibility study to discover potential technical solution, and the determination of system requirements.

It is at this early stage in the life cycle that major decisions are made relative to adapting a specific design approach and technology application, which has a great impact on the life-cycle cost of a product. In this phase, the designer addresses the fundamental question of whether to proceed with the selected concept. It is evident that there is no benefit or future in spending any more time and resource attempting to achieve an unrealistic objective. Some revolutionary concepts initially seem attractable, but when it comes to the reality, it is found to be too imaginary. Feasibility study distinguishes between a creative design concept and an imaginary idea. Feasibility evaluation determines the degree to which each concept alternative satisfies design criteria.

In this phase, the designer addresses the fundamental question of whether to proceed with the selected concept. Feasibility study distinguishes between a creative design concept and an imaginary idea. Feasibility evaluation determines the degree to which each concept alternative satisfies design criteria.

In the feasibility analysis, the answers to the following two questions are sought: (1) Are the goals achievable?; or are the objectives realistic?; or are the design requirements meetable? and (2) Is the current design concept feasible? If the answer to the first question is no, the design goal and objectives, and design requirements must be changed. Hence, no matter where is the source of design requirements; either direct customer order or market analysis; they must be changed.

1.7   DESIGN GROUPS

An aircraft chief designer should be capable of covering and handling a broad spectrum of activities. Thus, an aircraft chief designer should have years of experiences, be knowledgeable of management techniques, and preferably have full expertise and background in the area of "flight dynamics." The chief designer has a great responsibility in planning, coordination, and conducting formal design reviews. He/she must also monitor and review aircraft system test and evaluation activities, as well as coordinating all formal design changes and modifications for improvement. The organization must be such that facilitate the flow of information and technical data among various design departments. The design organization must allow the chief designer to initiate and establish the necessary ongoing liaison activities throughout the design cycle.

A primary building block is organizational patterns is the functional approach, which involves the grouping of functional specialties or disciplines into separately identifiable entities. The intent is to perform similar work within one organizational group. Thus, the same organizational group will accomplish the same type of work for all ongoing projects on a concurrent basis. The ultimate objective is to establish a team approach, with the appropriate communications, enabling the application of concurrent engineering methods throughout.

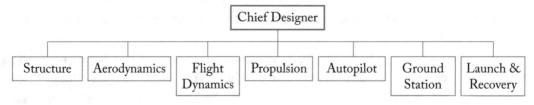

Figure 1.2: UAV main design groups.

There are two main approaches to handle the design activities and establishing design groups: (1) design groups based on aircraft components, and (2) design groups based on expertise (Figure 1.2). If the approach of groups based on aircraft components is selected, the chief designer must establish the following teams: (1) wing design team, (2) tail design team, (3) fuselage design team, (4) propulsion system design team, (5) landing gear design team, (6) autopilot design team, (7) ground station design team, and (8) launch and recovery design team. The ninth team is established for documentation, and drafting. There are various advantages and disadvantages for each of the two planning approaches in terms of ease of management, speed of communication, efficiency, and similarity of tasks. However, if the project is large, such as the design of a large transport aircraft, both groupings could be applied simultaneously.

1.8   DESIGN PROCESS

UAV Design is an iterative process which involves synthesis, analysis, and evaluation. Figure 1.3 demonstrates the design process block diagram. Design (i.e., synthesis) is the creative process of putting known things together into new and more useful combinations. Analysis refers to the process of predicting the performance or behavior of a design candidate. Evaluation is the process of performance calculation and comparing the predicted performance of each feasible design candidate to determine the deficiencies. A design process requires both integration and iteration. There is an interrelationship between synthesis, analysis, and evaluation. Two main groups of design activities are: (1) problem solving through mathematical calculations, and (2) choosing a preferred one among alternatives.

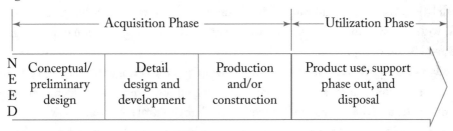

Figure 1.3: The UAV life-cycle.

In general, design considerations are the full range of attributes and characteristics that could be exhibited by an engineered system, product, or structure. These interest both the producer and the customer. Design-dependent parameters are attributes and/or characteristics inherent in the design to be predicted or estimated (e.g., weight, design life, reliability, producibility, maintainability, and disposability). These are a subset of the design considerations for which the producer is primarily responsible. On the other hand, design-independent parameters are factors external to the design that must be estimated and forecasted for use in design evaluation (e.g., fuel cost per gallon, interest rates, labor rates, and material cost per pound). These depend upon the production and operating environment of the UAV.

A goal statement is a brief, general, and ideal response to the need statement. The objectives are quantifiable expectations of performance which identify those performance characteristics of a design that are of most interest to the customer. Restrictions of function of form are called constraints; they limit our freedom to design.

1.9   SYSTEMS ENGINEERING APPROACH

Complex UAV systems, due to the high cost and the risks associated with their development become a prime candidate for the adoption of systems engineering methodologies. The UAV conceptual design process has been documented in many texts, and the interdisciplinary nature of the system is immediately apparent. A successful configuration designer needs not only a good understanding of design, but also systems engineering approach. A competitive configuration design manager must have a clear idea of the concepts, methodologies, models, and tools needed to understand and apply systems engineering to UAV systems.

The design of a UAV begins with the requirements definition and extends through functional analysis and allocation, design synthesis and evaluation, and finally validation. An optimized UAV, with a minimum of undesirable side effects, requires the application of an integrated life-cycle oriented "system" approach. The design of the configuration for the UAV begins with the requirements definition and extends through functional analysis and allocation, design synthesis and evaluation, and finally validation. Operations and support needs must be accounted for in this process. An optimized UAV, with a minimum of undesirable side effects, requires the application of an integrated life-cycle oriented "system" approach.

The design of the UAV subsystems plays a crucial role in the configuration design and their operation. These subsystems turn an aerodynamically shaped structure into a living, breathing, unmanned flying machine. These subsystems include the: flight control subsystem, power transmission subsystem, fuel subsystem, structures, propulsion, aerodynamics, and landing gear. In the early stages of a conceptual or a preliminary design these subsystems must initially be defined, and their impact must be incorporated into the design layout, weight analysis, performance calculations, and cost benefits analysis.

A UAV is a system composed of a set of interrelated components working together toward some common objective or purpose. Primary objectives include safe flight achieved at a low cost. Every system is made up of components or subsystems, and any subsystem can be broken down into smaller components. For example, in an air transportation system, the UAV, terminal, ground support equipment, and controls are all subsystems. The UAV life-cycle is illustrated in Figure 1.3.

A UAV must feature product competitiveness, otherwise, the producer and designer may not survive in the world marketplace. Product competitiveness is desired by UAV producers worldwide. Accordingly, the systems engineering challenge is to bring products and systems into being that meet the mission expectations cost-effectively. Because of intensifying international competition, UAV producers are seeking ways to gain sustainable competitive advantages in the marketplace.

It is essential that UAV designers be sensitive to utilization outcomes during the early stages of UAV design and development. They also need to conduct life-cycle engineering as early as possible in the design process. Fundamental to the application of systems engineering is an understanding of the system life-cycle process illustrated in Figure 1.3. It must simultaneously embrace the life cycle of the manufacturing process, the life cycle of the maintenance and support capability, and the life cycle of the phase-out and disposal process.

The requirements need for a specific new UAV first comes into focus during the conceptual design process. It is this recognition that initiates the UAV conceptual design process to meet these needs. Then, during the conceptual design of the UAV, consideration should simultaneously be given to its production and support. This gives rise to a parallel life cycle for bringing a manufacturing capability into being.

Traditional UAV configuration design attempts to achieve improved performance and reduced operating costs by minimizing maximum takeoff weight. From the point of view of a UAV customer, however, this method does not guarantee the optimality of a UAV program. Multidisciplinary design optimization (MDO) is an important part of the UAV configuration design process. It first discusses the design parameters, constraints, objectives functions, and criteria and then UAV configuration classifications. Then the relationship between each major design option and the design requirements are evaluated. Then the systems engineering principals are presented. At the end, systems engineering approach is applied in the optimization of the UAV configuration design and a new configuration design optimization methodology is introduced.

The design of a UAV within the system life-cycle context is different from the design just to meet a set of performance or stability requirements. Life-cycle focused design is simultaneously responsive to customer needs and to life-cycle outcomes. The design of the UAV should not only transform a need into a UAV/system configuration, but should ensure the UAV's compatibility with related physical and functional requirements. Further, it should consider operational outcomes expressed as safety, producibility, affordability, reliability, maintainability, usability, supportability, serviceability, disposability, and others, as well as the requirements on performance, stability, control, and effectiveness.

An essential technical activity within this process is that of evaluation. Evaluation must be inherent within the systems engineering process and must be invoked regularly as the system design activity progresses. However, systems evaluation should not proceed without guidance from customer requirements and specific system design criteria. When conducted with full recognition of design criteria, evaluation is the assurance of continuous design improvement. There are a number of phases through which the system design and development process must invariably pass. Foremost among them is the identification of the customer related need and, from that need, the determination of what the system is to do. This is followed by a feasibility analysis to discover potential technical solutions, the determination of system requirements, the design and development of system components, the construction of a prototype, and/or engineering model, and the valida-tion of system design through test and evaluation. The system (e.g., UAV) design process includes four major phases: (1) Conceptual Design, (2) Preliminary Design, (3) Detail Design, and (4) Test and Evaluation. The four phases of the integrated design of a UAV are summarized in Figure 1.4. Sections 1.10–1.13 present the details of these design phases.

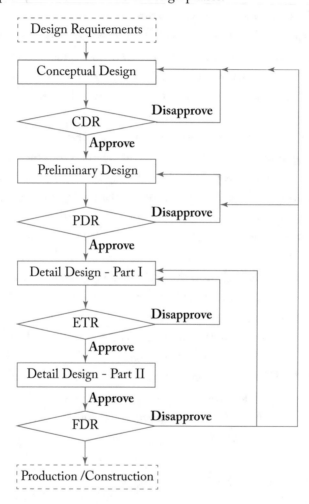

Figure 1.4: Design process and formal design reviews.

In the conceptual design phase, the UAV will be designed in concept without the precise calculations. In another word, almost all parameters are determined based on a decision making process and a selection technique. On the other hand, the preliminary design phase tends to employ the outcomes of a calculation procedure. As the name implies, in the preliminary design phase, the parameters that are determined are not final and will be altered later. In addition, in this phase, parameters are essential and will directly influence the entire detail design phase. Therefore the ultimate care must be taken to insure the accuracy of the results of the preliminary design phase. In the detail design phase, the technical parameters of all components (e.g., wing, fuselage, tail, landing gear (LG), and engine) including geometry are calculated and finalized.

1.10   CONCEPTUAL DESIGN

Throughout the conceptual system design phase (commencing with the need analysis), one of the major objectives is to develop and define the specific design-to requirements for the system as an entry. The results from these activities are combined, integrated, and included in a system specification. This specification constitutes the top "technical-requirements" document that provides overall guidance for system design from the beginning. Conceptual design is the first and most important phase of the UAV system design and development process. It is an early and high-level life cycle activity with potential to establish, commit, and otherwise predetermine the function, form, cost, and development schedule of the desired UAV system. The identification of a problem and associated definition of need provides a valid and appropriate starting point for design at the conceptual level.

Selection of a path forward for the design and development of a preferred system configuration, which will ultimately be responsive to the identified customer requirement, is a major responsibility of conceptual design. Establishing this early foundation, as well as requiring the initial planning and evaluation of a spectrum of technologies, is a critical first step in the implementation of the systems engineering process. Systems engineering, from an organizational perspective, should take the lead in the definition of system requirements from the beginning and address them from a total integrated life-cycle perspective.

As the name implies, the UAV conceptual design phase is the UAV design at the concept level. At this stage, the general design requirements are entered in a process to generate a satisfactory configuration. The primary tool in this stage of design is the "selection." Although there are variety of evaluation and analysis, but there are no much calculation. The past design experience plays a crucial role in the success of this phase. Hence, the members of conceptual design phase team must be the most experienced engineers of the corporation. Figure 1.5 illustrates the major activities which are practiced in the UAV conceptual design phase. The fundamental output of this phase is an approximate three-view of the UAV that represents the UAV configuration.

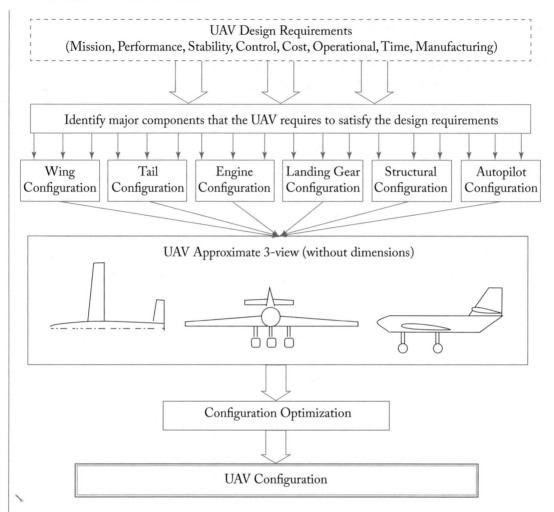

Figure 1.5: UAV conceptual design.

A UAV comprised of several major components. It mainly includes wing, horizontal tail, vertical tail, fuselage, propulsion system, landing gear, control surfaces, and autopilot. In order to make a decision about the configuration of each UAV component, the designer must be fully aware of the function of each component. Each UAV component has inter-relationships with other components and interferes with the functions of other components. The above six components are assumed to be the fundamental components of an air vehicle. However, there are other components in a UAV that are not assumed here as a major one. The roles of those components are described in the later sections whenever they are mentioned. Table 1.4 illustrates a summary of UAV major components and their functions. This table also shows the secondary roles and the major areas of influence of each UAV component. The table also specifies the design requirements that are affected by each component.

Table 1.5 illustrates a summary of configuration alternatives for UAV major components. In this table, various alternatives for wing, horizontal tail, vertical tail, fuselage, engine, landing gear, control surfaces, and automatic control system or autopilot are counted. An autopilot tends to function in three areas of guidance, navigation and control. More details are given in the detail design phase section. For each component, the UAV designer must select one alternative which satisfies the design requirements at an optimal condition. The selection process is based on a trade-off analysis with comparing all pros and cons in conjunction with other components.

Table 1.4: UAV major components and their functions			
No	Component	Primary Function	Major Areas of Influence
1	Fuselage	Payload accommodations	UAV performance, longitudinal stability, lateral stability, cost
2	Wing	Generation of lift	UAV performance, lateral stability
3	Horizontal tail	Longitudinal stability	Longitudinal trim and control
4	Vertical tail	Directional stability	Directional trim and control, stealth
5	Engine	Generation of thrust	UAV performance, stealth, cost, control
6	Landing gear	Facilitate take-off and landing	UAV performance, stealth, cost
7	Control surfaces	Control	Maneuverability, cost
8	Autopilot	Control, guidance, and navigation	Maneuverability, stability, cost, flight safety
9	Ground station	Control and guide the UAV from the ground	Autonomy, flight safety
10	Launch and recovery	Launching and recovering the UAV	Propulsion, structure, launcher, recovery system

Table 1.5: UAV major components with design alternatives

No	Component	Configuration Alternatives
1	Fuselage	- Geometry: lofting, cross section - Internal arrangement - What to accommodate (e.g., fuel, engine, and landing gear)?
2	Wing	- Type: swept, tapered, dihedral; - Location: low-wing, mid-wing, high wing, parasol - High lift device: flap, slot, slat - Attachment: cantilever, strut-braced
3	Horizontal tail	- Type: conventional, T-tail, H-tail, V-tail, inverted V - Installation: fixed, moving, adjustable - Location: aft tail, canard, three surfaces
4	Vertical tail	Single, twin, three VT, V-tail
5	Engine	- Type: turbofan, turbojet, turboprop, piston-prop, rocket - Location: (e.g., under fuselage, under wing, beside fuselage) - Number of engines
6	Landing gear	- Type: fixed, retractable, partially retractable - Location: (e.g., nose, tail, multi)
7	Control surfaces	Separate vs. all moving tail, reversible vs. irreversible, conventional vs. non-conventional (e.g., elevon, ruddervator)
8	Autopilot	- UAV: Linear model, nonlinear model - Control subsystem: PID, gain scheduling, optimal, QFT, robust, adaptive, intelligent - Guidance subsystem: Proportional Navigation Guidance, Line Of Sight, Command Guidance, three point, Lead - Navigation subsystem: Inertial navigation (Strap down, stable platform), GPS
9	Launch and recovery	HTOL, ground launcher, net recovery, belly landing

In order to facilitate the conceptual design process, Table 1.6 shows the relationship between UAV major components and the design requirements. The third column in Table 1.6 clarifies the UAV component which affected most; or major design parameter by a design requirement. Every design requirement will normally affects more than one component, but we only consider the component that is influenced most. For example, the payload requirement, range and endurance will affect maximum take-off weight, maximum take-off weight, engine selection, fuselage design, and flight cost. The influence of payload weight is different than payload volume. Thus, for optimization purpose, the designer must know exactly payload weight and its volume. On the other hand, if the payload can be divided into smaller pieces, the design constraints by the payload are easier to handle. Furthermore, the other performance parameters (e.g., maximum speed, stall speed, rate of climb, take-off run, ceiling) will affect the wing area and engine power (or thrust).

Table 1.6: Relationship between UAV major components and design requirements		
No	Design Requirements	UAV Component that Affected Most, or Major Design Parameter
1	Payload (weight) requirements	Maximum take-off weight
	Payload (volume) requirements	Fuselage
2	Performance Requirements (range and endurance)	Maximum take-off weight
3	Performance requirements (maximum speed, Rate of climb, take-off run, stall speed, ceiling, and turn performance)	Engine; landing gear; and wing
4	Stability requirements	Horizontal tail and vertical tail
5	Controllability requirements	Control surfaces (elevator, aileron, rudder), autopilot
6	Autonomy requirements	Center of gravity, autopilot, ground station
7	Airworthiness requirements	Minimum requirements, autopilot
8	Cost requirements	Materials; engine; weight, etc.
9	Timing requirements	Configuration optimality
10	Trajectory requirements	Autopilot

In order to select the best UAV configuration, a trade-off analysis must be established. Many different trade-offs are possible as the UAV design progresses. Decisions must be made regarding the evaluation and selection of appropriate components, subsystems, possible degree of automation, commercial off-the-shelf parts, various maintenance and support policies, and so on. Later in the design cycle, there may be alternative engineering materials, alternative manufacturing processes, alternative factory maintenance plans, alternative logistic support structures, and alternative methods of material phase-out, recycling, and/or disposal.

The UAV designer must first define the problem statement, identify the design criteria or measures against which the various alternative configurations will be evaluated, the evaluation process, acquire the necessary input data, evaluate each of the candidate under consideration, perform a sensitivity analysis to identify potential areas of risk, and finally recommend a preferred approach. Only the depth of the analysis and evaluation effort will vary, depending on the nature of the component.

Trade-off analysis involves synthesis which refers to the combining and structuring of components to create a UAV system configuration. Synthesis is design. Initially, synthesis is used in the development of preliminary concepts and to establish relationships among various components of the UAV. Later, when sufficient functional definition and decomposition have occurred, synthesis is used to further define "hows" at a lower level. Synthesis involves the creation of a configuration that could be representative of the form that the UAV will ultimately take (although a final configuration should not be assumed at this early point in the design process).

One of the most effective techniques in trade-off studies is multidisciplinary design optimization. Researchers in academia, industry, and government continue to advance Multidisciplinary Design Optimization (MDO) and its application to practical problems of industry relevance. Multidisciplinary design optimization is a field of engineering that uses optimization methods to solve design problems incorporating a number of disciplines. Multidisciplinary design optimization allows designers to incorporate all relevant disciplines simultaneously. The optimum solution of a simultaneous problem is superior to the design found by optimizing each discipline sequentially, since it can exploit the interactions between the disciplines. However, including all disciplines simultaneously significantly increases the complexity of the design problem.

1.11   PRELIMINARY DESIGN

Four fundamental UAV parameters are determined during the preliminary design phase: (1) UAV maximum take-off weight (W_{TO}), (2) wing reference area (S), (3) engine thrust (T) or engine power (P), and (4) autopilot preliminary calculations. Hence, four primary UAV parameters of W_{TO}, S, T (or P), and several autopilot data are the output of the preliminary design phase. These four parameters will govern the UAV size, the manufacturing cost, and the complexity of calculation. If during the conceptual design phase, a jet engine is selected, the engine thrust is calculated during this phase. But, if during the conceptual design phase, a prop-driven engine is selected, the engine power is calculated during this phase. A few other non-important UAV parameters such as UAV zero-lift drag coefficient and UAV maximum lift coefficient are estimated in this phase too.

Figure 1.6 illustrates a summary of the preliminary design process. The preliminary design phase is performed in three steps: (1) estimate UAV maximum take-off weight; (2) determine wing area and engine thrust (or power) simultaneously; and (3) autopilot preliminary calculations.

In this design phase, three design techniques are employed. First, a technique based on the statistics is used to determine UAV maximum take-off weight. The design requirements which are used in this technique are flight mission, payload weight, range, and endurance.

Next, another technique is employed based on the UAV performance requirements (such as stall speed, maximum speed, range, rate of climb, and take-off run) to determine the wing area and the engine thrust (or engine power). This technique is sometime referred to as the matching plot or matching chart, due to its graphical nature and initial sizing. The principles of the matching plot technique are originally introduced in a NASA technical report and they were later developed by Sadraey [37]. The technique is further developed by the author in his new book on UAV design that is under publication.

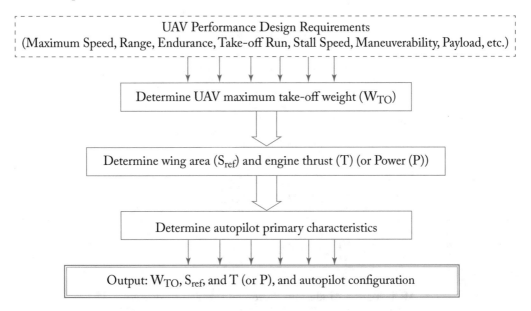

Figure 1.6: Preliminary design procedure.

In general, the first technique is not accurate (in fact, it is an estimation) and the approach may carry some inaccuracies, while the second technique is very accurate and the results are reliable. Due to the length of the book, the details of these three techniques have not been discussed in details here in this section. It is assumed that the reader is aware of these techniques which are practiced in many institutions.

1.12   DETAIL DESIGN

The design of the UAV subsystems and components plays a crucial role in the success of the flight operations. These subsystems turn an aerodynamically shaped structure into a living, breathing, unmanned flying machine. These subsystems include the: wing, tail, fuselage, flight control subsystem, power transmission subsystem, fuel subsystem, structures, propulsion, landing gear, and autopilot. In the early stages of a conceptual or a preliminary design phase, these subsystems must initially be defined, and their impact must be incorporated into the design layout, weight analysis, performance/stability calculations, and cost benefits analysis. In this section, the detail design phase of a UAV is presented.

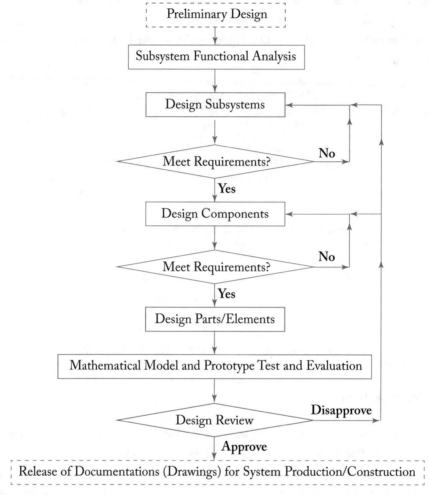

Figure 1.7: Detail design sequence.

As the name implies, in the detail design phase, the details of parameters of all major components (Figure 1.7) of a UAV is determined. This phase is established based on the results of conceptual design phase and preliminary design phase. Recall that the UAV configuration has been determined in the conceptual design phase and wing area, engine thrust, and autopilot major features have been set in preliminary design phase. The parameters of wing, horizontal tail, vertical tail, fuselage, landing gear, engine, subsystems, and autopilot must be determined in this last design phase. To compare three design phases, the detail design phase contains a huge amount of calculations and a large mathematical operation compared with other two design phases. If the total length of a UAV design is considered to be one year, about ten months is spent on the detail design phase.

This phase is an iterative operation in its nature. In general, there are four design feedbacks in the detail design phase. Figure 1.4 illustrates the relationships between detail design and design feedbacks. Four feedbacks in the detail design phase are: (1) performance evaluation, (2) stability analysis, (3) controllability analysis, and (4) flight simulation. The UAV performance evaluation includes the determination of UAV zero-lift drag coefficient. The stability analysis requires the component weight estimation plus the determination of UAV center of gravity (cg). In the controllability analysis operation, the control surfaces (e.g., elevator, aileron, and rudder) must be designed. When the autopilot is designed, the UAV flight needs to be simulated to assure the flight success.

As the name implies, each feedback is performed to compare the output with the input and correct the design to reach the design goal. If the performance requirements are not achieved, the design of several components, such as engine and wing, might be changed. If the stability requirements are not met, the design of several components, such as wing, horizontal tail and vertical tail could be changed. If the controllability evaluation indicates that the UAV does not meet controllability requirements, control surfaces and even engine must be redesigned. In case that both stability requirements and controllability requirements were not met, the several components must be moved to change the cg location.

In some instances, this deficiency may lead to a major variation in the UAV configuration, which means the designer needs to return to the conceptual design phase and begin the correction from the beginning. The deviation of the UAV from trajectory during flight simulation necessitates a change in autopilot design.

1.13   DESIGN REVIEW, EVALUATION, AND FEEDBACK

In each major design phase (conceptual, preliminary, and detail), an evaluation should be conducted to review the design and to ensure that the design is acceptable at that point before proceeding with the next stage. There is a series of formal design reviews conducted at specific times in the overall system development process. An essential technical activity within the design process is that of evaluation. Evaluation must be inherent within the systems engineering process and must be invoked regularly as the system design activity progresses. When conducted with full recognition of design criteria, evaluation is the assurance of continuous design improvement. The evaluation process includes both the informal day-to-day project coordination and data review, and the formal design review.

The purpose of conducting any type of review is to assess if (and how well) the design configuration, as envisioned at the time, is in compliance with the initially specified quantitative and qualitative requirements. A design review provides a formalized check of the proposed system design with respect to specification requirements. In principle, the specific types, titles, and scheduling of these formal reviews vary from one design project to the next. The following four main formal design reviews are recommended for a design project.

1. Conceptual Design Review (CDR)

2. Preliminary Design Review (PDR)

3. Evaluation and Test Review (ETR)

4. Critical (Final) Design Review (FDR)

Figure 1.6 shows the position of each design review in the overall design process. Design reviews are usually scheduled before each major design phase. The CDR is usually scheduled toward the end of the conceptual design phase and prior to entering the preliminary design phase of the program. The purpose of conceptual design review (CDR) is to formally and logically cover the proposed design from the system standpoint. The preliminary design review is usually scheduled toward the end of the preliminary design phase and prior to entering the detail design phase. The critical design review (FDR) is usually scheduled after the completion of the detail design phase and prior to entering the production phase.

The evaluation and test review is usually scheduled somewhere in the middle of the detail design phase and prior to production phase. The ETR accomplishes two major tasks: (1) finding and fixing any design problems and the subsystem/component level, and then (2) verifying and documenting the system capabilities for government certification or customer acceptance. The ETR can range from the test of a single new system for an existing system to the complete development and certification of a new system.

1.14   QUESTIONS

1. What are the five terms which are currently used for unmanned aircraft?

2. What are the primary design requirements for a UAV?

3. Describe features of a Tier II UAV in the Air Forces.

4. Describe the features of a micro UAV.

5. What is the main objective for the feasibility study?

6. What is the size range for mini UAVs?

7. What do MALE and HALE stand for?

8. What is the operating altitudes for HALE UAVs?

9. What is the endurance range for MALE UAVs?

10. What are the wingspan and MTOW of Global Hawk?

11. What are the cruise speed and endurance for Predator (RQ-1A)?

12. What was the major setback during Phase II flight testing of the Global Hawk on March 29, 1999? What was the reason behind that?

13. Describe the fundamentals of systems engineering approach in UAV design.

14. What are the main four formal design reviews?

15. What are the UAV main design groups?

16. Describe conceptual design phase.

17. Describe main outputs of the preliminary design.

18. Describe process of detail design.

19. Describe trade-off analysis process.

20. From systems engineering approach, what are the main design phases?

CHAPTER 2

Design Disciplines

2.1 INTRODUCTION

There are several design disciplines which work in parallel within a UAV design project. Some examples are: (1) aerodynamic design, (2) structural design, (3) propulsion system design, (4) power transmission system design, (5) mechanical system design, and (6) control surfaces design, (7) ground station, and (8) launch and recovery system. This chapter briefly covers the first six topics disciplines; the other two are presented in Chapters 8 and 9, respectively. Due to the limited volume of the book, only the basic fundamentals are presented. The interested reader should refer to Sadraey [37] for more details. Table 2.1 shows the UAV major components and their primary functions.

No	Component	Primary function	Major areas of influence
\multicolumn{4}{l}{Table 2.1: UAV vehicle major components and their functions}
1	Fuselage	Accommodations (Payload, systems)	UAV performance, longitudinal stability, lateral stability, cost, stability
2	Wing	Generation of lift	UAV performance, lateral stability
3	Horizontal/Vertical tail	Longitudinal/Directional stability	Longitudinal/Directional trim and control, stealth
4	Payload	Sense, measure, release/drop a store	UAV weight, drag, performance, power consumption
5	Engine	Generation of thrust/power	UAV performance, stealth, cost, control
6	Landing gear	Facilitate take-off and landing	UAV performance, stealth, cost
7	Control surfaces	Create rolling, pitching, and yawing moment	Maneuverability, cost
8	Autopilot	Control, guidance, and navigation	Maneuverability, stability, cost, flight safety

In a UAV design process, some UAV parameters must be minimized (e.g., weight), while some other variables must be maximized within constraints (e.g., range, endurance, maximum speed, and ceiling), and also others must be evaluated to ensure that they are acceptable. The optimization process must be accomplished through a systems engineering approach. In some cases, the design of the UAV may impose slight to considerable changes to the UAV mission during the conceptual design process. The strong relationship between the analysis and the influencing parameters allow definite, traceable relationships to be constructed. In the case of a UAV design, the major parameters are derived almost completely from operational and performance requirements.

It is clear that some steps may be moved along with regard to the UAV mission, design team members, past design experiences, design facility, and manufacturing technologies. As it is observed, the design process is truly an iterative process and there are several modification steps to satisfy all design requirements. An important feature of the design process is the lessons learned in the past. The lesson will be utilized in improving the next generation, for instance, the major setback during Phase II flight testing of Global Hawk (Figure 1.1) was the destruction of air vehicle 2 on March 29, 1999, during the program's 18th sortie. The loss of air vehicle 2 and its payload was estimated at $45 million. Of more importance, however, was the fact that the program lost its only integrated sensor suite. The crash was due to a lack of proper frequency coordination between the Nellis Air Force Base and EAFB flight test ranges. Essentially, Nellis officials who were testing systems in preparation for Global Hawk's first planned D&E exercise were unaware that Global Hawk was flying over China Lake Naval Air Weapons Station, which is within EAFB's area of responsibility. Thus, many changes have been applied in the design of Northrop Grumman RQ-4B Global Hawk as compared with RQ-4A. For instance, the Northrop Grumman RQ-4B Global Hawk has a 50% payload increase, larger wingspan (130.9 ft) and longer fuselage (47.6 ft), and new generator to provide 150% more electrical output.

The integration of system engineering principles with the analysis-driven UAV design process demonstrates that a higher level of integrated vehicle can be attained; identifying the requirements/functional/physical interfaces and the complimentary technical interactions which are promoted by this design process. The details of conceptual design phase, preliminary design phase, and detail design phase were introduced in Chapter 1. In this chapter, design activities in several underlying disciplines are provided.

2.2   AERODYNAMIC DESIGN

The primary aerodynamic function of the UAV components (e.g., wing) is to generate sufficient lift force or simply lift (L). However, they have two other productions, namely drag force or drag (D) and nose-down pitching moment (M). While a UAV designer is looking to maximize the lift, the other two (drag and pitching moment) must be minimized. In fact, a wing is considered as a lifting surface that lift is produced due to the pressure difference between lower and upper surfaces. Aerodynamics textbooks are a good source to consult for information about mathematical techniques for calculating the pressure distribution over the wing and for determining the flow variables.

During the aerodynamic design process, several parameters must be determined. For instance, for a wing, they are as follows: (1) wing reference (or planform) area, (2) number of the wings, (3) vertical position relative to the fuselage (high, mid, or low wing), (4) horizontal position relative to the fuselage, (5) cross-section (or airfoil), (6) aspect ratio (AR), (7) taper ratio (λ), (8) tip chord (Ct), (9) root chord (Cr), (10) mean Aerodynamic Chord (MAC or C), (11) span (b), (12) twist angle (or washout) (α_t), (13) sweep angle (Λ), (14) dihedral angle (Γ), (15) incidence (i_w) (or setting angle, α_{set}), (16) high-lift devices such as flap, (17) aileron, and (18) other wing accessories.

One of the necessary tools in the wing design process is an aerodynamic technique to calculate wing lift, wing drag, and wing pitching moment. With the progress of the science of aerodynamics, there are variety of techniques and tools to accomplish this time consuming job. A variety of tools and software based on aerodynamics and numerical methods have been developed in the past decades. The CFD[1] Software based on the solution of Navier-Stokes equations, vortex lattice method, thin airfoil theory, and circulation are available in the market. The application of such software packages, which is expensive and time-consuming, at this early stage of wing design seems unnecessary.

Wing is a three-dimensional component, while the airfoil is a two-dimensional section. Because of the airfoil section, two other outputs of the airfoil, and consequently the wing, are drag and pitching moment. The wing may have a constant or a non-constant cross-section across the wing. There are two ways to determine the wing airfoil section, the airfoil design and the airfoil selection. The design of the airfoil is a complex and time consuming process and needs expertise in fundamentals of aerodynamics at graduate level. Since the airfoil needs to be verified by testing it in a wind tunnel, it is expensive too.

Selecting an airfoil is a part of the overall wing design. Selection of an airfoil for a wing begins with the clear statement of the flight requirements. For instance, a subsonic flight design requirements are very much different from a supersonic flight design objectives. On the other hand, flight in the transonic region requires a special airfoil that meets Mach divergence requirements. The designer must also consider other requirements such as airworthiness, structural, manufacturability, and cost requirements. In general, the following are the criteria to select an airfoil for a wing with a collection of design requirements:

1. the airfoil with the highest maximum lift coefficient ($C_{l_{max}}$);

2. the airfoil with the proper ideal or design lift coefficient (C_{l_d} or C_{l_i});

3. the airfoil with the lowest minimum drag coefficient ($C_{d_{min}}$);

4. the airfoil with the highest lift-to-drag ratio (($C_l/C_d)_{max}$);

---

[1] Computational Fluid Dynamics

5. the airfoil with the highest lift curve slope ($C_{l_{\alpha\max}}$);

6. the airfoil with the lowest (closest to zero; negative or positive) pitching moment coefficient (C_m);

7. the proper stall quality in the stall region (the variation must be gentle, not sharp);

8. the airfoil must be structurally reinforceable. The airfoil should not that much thin that spars cannot be placed inside;

9. the airfoil must be such that the cross section is manufacturable;

10. the cost requirements must be considered; and

11. other design requirements must be considered. For instance, if the fuel tank has been designated to be places inside the wing inboard section, the airfoil must allow the sufficient space for this purpose.

12. If more than one airfoil is considered for a wing, the integration of two airfoils in one wing must be observed.

In designing the high lift device for a wing, the following parameters must be determined: (1) high lift device location along the span; (2) the type of high lift device; (3) high lift device chord (C_f); (4) high lift device span (b_f); and (5) high lift device maximum deflection (down) ($\delta_{f\max}$). For fundamentals of aerodynamics, please refer to references such as Anderson [44] and Shevell [45].

2.3   STRUCTURAL DESIGN

The structure of a conventional fixed-wing UAV consists of five principal units: fuselage, wings, horizontal tail, vertical tail, and control surfaces. The landing gear is also part of structure, but will be covered in Section 2.5. Engine pylon, engine inlet (for supersonic UAVs), fairings (and fillets), and landing gear bay doors are also assumed as part of aircraft structure. The primary functions of the structure is (1) to keep the aerodynamic shape of the UAV and (2) to carry the loads. Airframe structural components are constructed from a wide variety of materials. The earliest aircraft were constructed primarily of wood. Steel tubing and the most common material, aluminum, followed. Many newly certified aircraft are built from molded composite materials, such as glass/epoxy and carbon fiber.

Structural members of a fuselage mainly include stringers, longerons, bulkheads, and skin. The structural members in a wing/tail are spar, rib, stiffeners, and skin. The fuselage/wing/tails skin can be made from a variety of materials, ranging from impregnated fabric to plywood, aluminum, or composites. Under the skin and attached to the structural components are the many components that support airframe function. The entire airframe and its components are joined by rivets, bolts, screws, and other fasteners. Welding, adhesives, and special bonding techniques are also employed.

The most common form of UAV structure is semi-monologue (single shell) which implies that the skin is stressed/reinforced. The structural members are designed to carry the flight loads or to handle stress without failure. In designing the structure, every square inch of wing and fuselage, must be considered in relation to the physical characteristics of the material of which it is made. Every part of the structure be planned to carry the load which is applied on it.

The structural designer will determine flight loads, calculate stresses, and design structural elements such as to allow the UAV components to perform their aerodynamic functions efficiently. This goal will be considered simultaneously with the objective of the lowest structural weight. The most common tool in structural analysis is the use of finite element methods (FEM). One of the earliest and the most well-known computer software is NASTRAN, developed by NASA in the mid-1960s. The stress analysis is the basic calculation to determine the safety factor. There are five major stresses to which structural members are subjected: (1) tension, (2) compression, (3) torsion, (4) shear, and (5) bending. A single member of the structure is often subjected to a combination of stresses.

Fuselage usually consists of frame assemblies, bulkheads, and formers. The skin is reinforced by longitudinal members called longeron. Often, wings/tails are of full cantilever design. In general, wing construction is based on one of three fundamental designs: (1) monospar, (2) multispar, and (3) box beam. Spars are the principal structural members of the wing. They correspond to the longeron of the fuselage. Spars run parallel to the lateral axis of the aircraft, from the fuselage toward the tip of the wing, and are usually attached to the fuselage by a beam, or a truss. Generally, a wing has two spars. One spar is usually located at the maximum thickness, and the other about two-thirds of the distance toward the wing's trailing edge (in front of flap/control-surface).

Honeycomb structured wing panels are often used in composite wings. Nacelles (i.e., pods) are streamlined enclosures used primarily to house the engine and its components. Engine mounts are also found in the nacelle. These are the assemblies to which the engine is fastened. They are usually constructed from chrome/molybdenum steel tubing in light UAV and forged chrome/nickel/molybdenum assemblies in larger UAVs. Cowling are the detachable panels covering those areas into which access must be gained regularly, such as the engine and its accessories. In the design of airframe, several factors such as ultimate load, aerodynamic loads (pressure distribution), weight loads (e.g., fuel and engine), weight distribution, gust load, load factor, propulsion loads, landing loads (e.g., brake), and aero-elasticity effects must be considered.

One of the design requirements for some military UAVs is stealth. In the concept of stealth, the three basic methods of minimizing the reflection of pulses back to a receptor are: (1) to manufacture appropriate areas of the UAV from radar-translucent material such as Kevlar or glass composite as used in radomes which house radar scanners; (2) to cover the external surfaces of the aircraft with RAM (radar absorptive material); and (3) to shape the aircraft externally to reflect radar pulses in a direction away from the transmitter. The acoustic (i.e., noise) wavelength (signature) range for detecting an air vehicle is 16 m–2 cm.

The operating flight loads limits on a UAV are usually presented in the form of a V-n diagram. Structural designers will construct this diagram with the cooperation of the flight dynamics group. The diagram will determine the structural failure areas, and area of structural damage/failure. The UAV should not be flown out of the flight envelope, since it is not safe for the structures. The UAV structural design is out of scope of this book, you may refer to references such as Megson [47] for more details.

2.4   PROPULSION SYSTEM DESIGN

A heavier-than-air craft (UAV) requires a propulsion system in order to have a sustained flight. Without a proper aero-engine or powerplant, a heavier-than-air vehicle can only glide for a short time. The contribution of a powerplant to an aircraft is to generate the most influential force in the aircraft performance; that is, the propulsive force or thrust. The secondary function of the propulsion system is to provide power/energy to other subsystems such as hydraulic system, electric system, pressure system, air conditioning system, and avionics. These subsystems rely on the engine power to operate.

Soon after the design requirements and constraints are identified and prioritized, the propulsion system designer will begin to select the type of engine. There are a number of engine types available in the market for flight operations. They include: electric (battery), solar-powered, piston-prop, turbojet, turbofan, turboprop, turboshaft, ramjet, and rocket engines.

Propeller is a means to convert the engine power to engine thrust. The aerodynamic equations and principles that govern the performance of a wing are generally applied to a propeller. Hence, the propeller may be called a rotating wing. It simply creates the lift (i.e., thrust) with the cost of drag. For this reason, the prop efficiency never could reach 100%. In a cruising flight and in the optimum angle of twist (the best propeller pitch), the propeller efficiency (η_P) may operate at around 75%–85%. With this in mind, a method to estimate the propeller diameter is presented. The propeller design to determine parameters such as the blade airfoil section and twist angle is out of scope of this text.

An important issue in designing UAV engine is the type of fuel consumed by the engine. There are various requirements for the fuel, such as its density, boiling and freezing temperatures. Two issues for fuel when an aircraft is flying at high altitudes are that it evaporates and it freezes. Ignoring these two significant problems may result in the mission failure.

In general, the job of an aircraft designer is to determine/design/select the following items: (1) select engine type, (2) select number of engines, (3) determine engine location, (4) select engine from manufacturers' catalogs, (5a) size propeller (if prop-driven engine), (5b) design inlet (if jet engine), (6) design engine installation, and (7) iterate and optimize. Items 5a and 5b are design activities for an aircraft designer which must be carried along and parallel with the engine design team. This is to emphasize that the aircraft designer has the final say in these two propulsion system parameters. MQ-1A Predator (Figure 5.5) A was equipped with a piston engine—max speed of 80 knot—while the MQ-9B Predator B is equipped with a turboprop engine—max speed of 260 knot.

2.5   LANDING GEAR DESIGN

Another aircraft major component that is needed to be designed is landing gear (undercarriage). The landing gear is the structure that supports an aircraft on the ground and allows it to taxi, take-off, and land. In fact, landing gear design tends to have several interferences with the aircraft structural design. In general, the following are the landing gear parameters that are to be determined: (1) type (e.g., nose gear (tricycle), tail gear, bicycle), (2) fixed (faired, or un-faired), or retractable, partially retractable, (3) height, (4) wheel base, (5) wheel track, (6) the distance between main gear and aircraft cg, (7) strut diameter, (8) tire sizing (diameter, width), (9) landing gear compartment if retracted, and (10) load on each strut.

The landing gear usually includes wheels, but some aircraft are equipped with skis for snow or float for water. In the case of a vertical take-off and landing aircraft such as a helicopter, wheels may be replaced with skids. The landing gear is divided into two sections: (1) the main gear or main wheel,[2] and (2) the secondary gear or secondary wheel. The main gear is the gear which is the closest to the aircraft center of gravity (cg).

The primary functions of a landing gear are as follows:

1. to keep the aircraft stable on the ground and during loading, unloading, and taxi;

2. to allow the aircraft to freely move and maneuver during taxing;

3. to provide a safe distance between other aircraft components such as wing and fuselage while the aircraft is on the ground position to prevent any damage by the ground contact;

4. to absorb the landing shocks during landing operation; and

5. to facilitate take-off by allowing aircraft acceleration and rotation with the lowest friction.

In order to allow for a landing gear to function effectively, the following design requirements are established:

---

[2]   The term "wheel" is often used to mean the entire wheel/brake/tire assembly.

1. ground clearance requirement,

2. steering requirement,

3. take-off rotation requirement,

4. tip back prevention requirement,

5. overturn prevention requirement,

6. touch-down requirement,

7. landing requirement,

8. static and dynamic load requirement,

9. aircraft structural integrity,

10. ground lateral stability,

11. low cost,

12. low weight,

13. maintainability, and

14. manufacturability.

In general, there are nine configurations for a landing gear as follows: (1) single main, (2) bicycle, (3) tail-gear, (4) tricycle or nose-gear, (5) quadricycle, (6) multi-bogey, (7) releasable rail, (8) skid, and (9) seaplane landing device. In order to select the best landing gear configuration, the designer must perform a trade-off study using a comparison table.

Another design aspect of the landing gear is to decide what to do with it after take-off operation. In general, there are four alternatives:

1. landing gear is released after take-off;

2. landing gear hangs underneath the aircraft (i.e., fixed);

3. landing gear is fully retracted inside aircraft (e.g., inside wing, or fuselage); or

4. landing gear is partially retracted inside aircraft.

In order to formulate the design requirements, the designer is encouraged to develop several equations and relations based on the numerical requirements and solve them simultaneously.

2.6   MECHANICAL/POWER TRANSMISSION SYSTEMS DESIGN

One of the UAV subsystems that has a great influence on the control surfaces' design is the power system. The power required to operate a UAV includes that to drive the control surfaces' deflection, which is usually accomplished either hydraulically or electrically. There are several power systems employed in UAVs; each has advantages and disadvantages. They are: (1) electro-mechanical, (2) pneumatic, (3) hydraulic, and (4) electro-hydrostatic. In this section, advantages and disadvantages these systems, plus their influence on the design of the control surfaces, are briefly discussed. Commands can be sent to actuate the control surfaces using either a fly-by-wire or a fly-by-optic system. Figure 2.1 demonstrates an Insitu ScanEagle UAV on a ground launcher; it is launched by the pneumatic power. This UAV has a maximum takeoff weight of 48.5 lb, a wing area of 14.7 ft^2, and a piston engine with maximum power of 1.5 hp.

Figure 2.1: Insitu ScanEagle on a ground launcher.

Numerous changes have been made in aircraft flight control systems. Initially, flight control systems were purely mechanical, which was ideal for smaller, slow-speed, low-performance aircraft because they were easy to maintain. However, more control-surface force is required in modern high performance airplanes. Thus, during the 20th century a hydraulic power boost system was added to the mechanical control. This modification maintained the direct mechanical linkage between the pilot and the control surface. As aircraft became larger, faster, and heavier, and had increased performance, they became harder to control because the pilot could not provide the necessary power to directly operate the control surfaces. Thus, the entire effort of moving the control surface had to be provided by the actuator. A stability augmentation system (SAS) was added to the hydraulic boosted mechanical regulator system to make the aircraft flyable under all flight configurations. Motion sensors were used to detect aircraft perturbations and to provide electric signals to a SAS computer, which, in turn, calculated the proper amount of servo actuator force required.

The new step in the evolution of flight control systems is the use of a fly-by-wire (FBW) control system. In this design, all autopilot commands are transmitted to the control-surface actuators through electric wires. Thus, all mechanical linkages from the control stick to the servo actuators are removed from the UAV. The FBW system provided the advantages of reduced weight, improved survivability, and decreased maintenance.

Originally the flight control computers were analog (such as the F-16 aircraft computers), but these have been replaced by digital computers. In addition, the controller consists of a digital computer which accepts the pilot commands and signals from the sensors (position and rate gyros) and accelerometers, and sends commands to the actuators. This is now referred to as a digital flight control system (DFCS). For 21st century aerospace vehicles the use of hydraulics has essentially been eliminated in favor of an all-electric system incorporating the use of digital computers.

Mechanical flight control systems are the most basic design. They are used in a variety of aircraft. Currently electro-mechanical systems are used in small UAV where the aerodynamic forces are not excessive. Pure mechanical flight control systems use a collection of mechanical parts such as rods, cables, and pulleys to transmit commanded movements to the control surfaces from the actuators.

The complexity and weight of a mechanical flight control system increases considerably with the size and performance of the airplane. Hydraulic power overcomes many of these limitations. A hydraulic flight control system has two parts: the mechanical circuit, and the hydraulic circuit. A control command causes the mechanical circuit to open the matching servo valves in the hydraulic circuit. The hydraulic circuit powers the actuators which then move the control surfaces. However, hydraulic systems can require long hydraulic lines which may need to be redundant. These lines can develop leaks. Electro-hydrostatic actuators use local hydraulic reservoirs and pumps to avoid the need for these long hydraulic lines. Replacing the mechanical or hydraulic linkages with electrical power lines and linkages can save weight and improve reliability.

Electronic fly-by-wire systems can respond more flexibly to changing aerodynamic conditions. Fully electronic systems would require less maintenance, whereas mechanical and hydraulic systems require lubrication, tension adjustments, leak checks, fluid changes, etc. Digital flight-control systems were able to incorporate "multi-mode" flight control laws with different modes, each optimized to enhance maneuverability and controllability for a particular phase of flight. Furthermore, placing such control circuitry between the operator and the aircraft could provide various safety systems. For example, the control system could prevent a stall, or limit a maneuver which could overstress the structure.

Fly-by-optics systems are sometimes used instead of fly-by-wire systems because they can transfer data at higher speeds, and are nearly immune to electromagnetic interference. In a simplistic view, the electrical cables of the fly-by-wire system are replaced by fiber optic cables.

2.7   CONTROL SURFACES DESIGN

Two primary prerequisites for a safe flight are stability and controllability. The controllability requirements will influence the design of control surfaces and create variety of design constraints. Flight stability is defined as the inherent tendency of an aircraft to oppose any input and return to original trim condition if disturbed. When the summation of all forces along each three axes, and summation of all moments about each three axes are zero, an aircraft is said to be in trim or equilibrium. In this case, aircraft will have a constant linear speed and/or a constant angular speed. Control is the process to change the aircraft flight condition from an initial trim point to a final or new trim point. This is performed mainly by autopilot through moving the control surfaces/engine throttle. The desired change is basically expressed with a reference to the time that takes to move from initial trim point to the final trim point (e.g., pitch rate and roll rate).

Maneuverability is profoundly significant for fighter aircraft and missiles and is a branch of controllability. Control systems should be designed with sufficient redundancy to achieve two orders of magnitude more reliability than some desired level. Aircraft controllability is a function of a number of factors including control surfaces.

In general, control surfaces may be broadly classified into two types: conventional, and non-conventional. Conventional control surfaces are divided into two main groups, primary control surfaces and secondary control surfaces. The primary control surfaces (Figure 2.3) that are in charge of control of flight route and usually in a conventional aircraft are the aileron, the elevator, and the rudder.

The primary control surfaces of aileron, elevator, and rudder are, respectively, utilized for lateral control, longitudinal control and directional control. However they also largely contribute to lateral trim, longitudinal trim and directional trim of aircraft. In the majority of aircraft configurations, lateral and directional motions are coupled; hence, the aileron is also affecting directional motion and rudder affects the lateral motion. Conventional primary control surfaces are like plain flap, but their applications are different. When control surfaces are deflected, cambers of their related lifting surfaces (wing, horizontal tail, or vertical tail) are changed. Thus, the deflection of a control varies the aerodynamic forces; and consequently a resultant moment will influence the aircraft motion.

To analyze the aircraft control, a coordinate axis system must be defined. There are four coordinate systems—earth-fixed, body-fixed, wind axis system, and stability axes system. Here, for the purpose of control, a body-fixed coordinate system is adopted; where there are three orthogonal axes which follow the right-hand rule. The x-axis is along the fuselage (body) centerline passing through the aircraft center of gravity; the y-axis is perpendicular to x-axis and to the right (from top-view); and z-axis is perpendicular to xy plane (i.e., downward). Figure 2.2 illustrates the convention for positive directions of axes of aircraft. Positive roll is defined as a clockwise rotation about x axis as seen from the pilot seat (when in cruise; right wing down, left wing up). Similarly, positive pitch is defined as a clockwise rotation about y axis as seen from the pilot seat (nose up). Finally, positive yaw is defined as a clockwise rotation about z axis as seen from the pilot seat (nose to right). Figure 2.3 illustrates the conventional control surfaces. These conventions are significant and are used to develop design techniques in this book. Figure 2.4 depicts the axes and positive rotations convention.

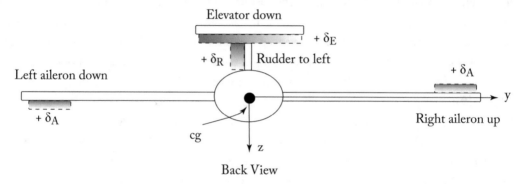

Figure 2.2: Convention for positive deflections of control surfaces.

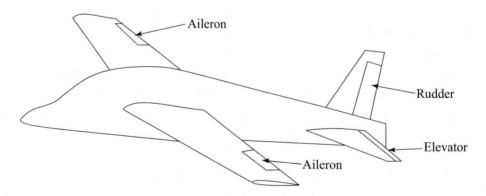

Figure 2.3: Primary control surfaces.

An aircraft is capable of performing various maneuvers and motions; they may be broadly classified into three main groups: (1) longitudinal motion, (2) lateral motion, and (3) directional motion.

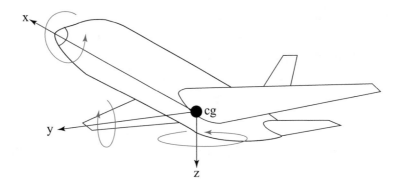

Figure 2.4: Axes and positive rotations convention

Figure 2.5 illustrates a flowchart that represents the control surfaces design process. In general, the design process begins with a trade-off study to establish a clear line between stability and controllability requirements and ends with optimization. During the trade-off study, two extreme limits of flying qualities are examined and the border line between stability and controllability is drawn. For instance a fighter can sacrifice the stability to achieve a higher controllability and maneuverability. Then, an automatic flight control system may be employed to augment the aircraft stability. In the case of a civil airliner, the safety is the utmost goal; so the stability is clearly favored over the controllability.

The first step in the design of control surfaces is to select the control surface configuration. The primary idea behind the design of flight control surfaces is to position them so that they function primarily as moment generators. They provide three types of rotational motions (roll, pitch, and yaw). A conventional configuration includes an elevator, aileron, and rudder. Variations to this classical configuration lead to some variations in the arrangements of these control surfaces. Table 2.2 represents several control surface configuration options. Some types of control surfaces are tied with particular aircraft configuration; they must be selected for specific aircraft configuration. Table 2.2 also illustrates a few aircraft examples.

The control surface configuration selection is a function of aircraft configuration (e.g., wing, tail, and engine), cost, performance, controllability power transmission, and operational requirements. The consequence of some aircraft configurations is to have a particular type of control surfaces. For instance, when a V-tail configuration is selected during the aircraft conceptual design phase, a ruddervator is the best candidate to control both yawing and pitching moments. Another example is when the designer decides to have a delta wing without aft tail. In such a case, an elevon is a great candidate to perform as a means of control power to control pitch rate and roll rate. The final decision on the control surface configuration will be the output of a trade-off study to balance and satisfy all design requirements in an optimum way. In general, unconventional control surfaces are more challenging to design, more complex to manufacture, and also harder to analyze. However, unconventional control surfaces are more efficient when a higher control power is required in a challenging design environment.

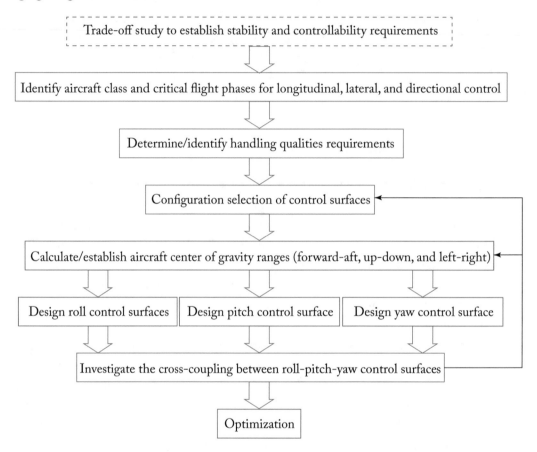

Figure 2.5: Control surfaces design process.

Based on performance requirements, a UAV will need multiple control surfaces such as elevators, ailerons, rudders, flaperons, ruddervators, and/or elevons. The number and the type of control surfaces depend on several factors such as the UAV's mission, cost, and controllability requirements. In this book, we concentrate on a conventional configuration that includes three control surfaces; namely the elevator, aileron, and rudder. The three coefficients of interest are: C_l (roll), C_m (pitch), and C_n (yaw). Data requirements may be included which can be used to minimize a particular objective as a function of control surface deflections. Other data requirements specific to constrained control allocation techniques include control minimum and maximum position limits and actuator rate limits.

No	Control Surface Configuration	Aircraft Configuration
\multicolumn{3}{l}{Table 2.2: Control surface configuration Options}		

No	Control Surface Configuration	Aircraft Configuration
1	Conventional (aileron, elevator, rudder)	Conventional (or canard replacing elevator)
2	All moving horizontal tail, rudder, aileron	Horizontal tail and elevator combined
3	All moving vertical tail, elevator, aileron	Vertical tail and rudder combined
4	Flaperon, Elevator, Rudder	Flap and aileron combined (e.g., X-29 and F-16 falcon)
5	Taileron, Rudder	All moving horizontal tail (elevator) and aileron combined (e.g., F-16 Falcon)
6	Elevon, Rudder (or equivalent)	Aileron and elevator combined (e.g., Dragon, F-117 Night Hawk, Space Shuttle)
7	Ruddervator, Aileron	V-tail (e.g., Global Hawk and Predator)
8	Drag-Rudder, Elevator, Aileron	No vertical tail (e.g., DarkStar)
9	Canardvator, Aileron	Elevator as part of canard, plus aileron
10	Four Control Surfaces	Cross (+ or ×) tail configuration (e.g., most missiles)
11	Aileron, Elevator (or equivalent), Split Rudder	No vertical tail. Aileron-like surfaces that is split into top and bottom sections (e.g., bomber B-2 Spirit)
12	Spoileron, Elevator, Rudder	Spoiler and aileron combined (e.g., Boeing B-52)
13	Thrust vector control	Augmented or no control surfaces, VTOL UAV

This data may also be dependent on other variables. As an example, many control laws have constraints implemented in software, determined by variables such as the dynamic pressure, which are imposed on the available surface deflections. For example, the commanded control actuator rate may need to change to maintain a constant aircraft rotation rate across the entire flight envelope, since hinge moments and other aerodynamic factors normally change with flight condition.

One flight control effector innovation is the use of thrust vectoring. One such flight control design is the F-15 ACTIVE vehicle which used thrust vectoring to produce pitch, roll, and yaw controlling moments. Other examples of such vehicles exist. One of the first operational aircraft to use thrust vectoring was the Harrier, flown by the British Royal Air Force, British Royal Navy, and the U.S. Marines. NASA research explored thrust vectoring at extremely high angle-of-attack on the High Alpha (angle-of-attack) Research Vehicle (HARV), a modified F-18. The F-16 Multi-Axis Thrust Vectoring (MATV) research program made significant contributions to understanding thrust vectoring design requirements and agility benefits. The X-31 research aircraft also demonstrated the benefits of thrust vectoring. The F-22 and the F-35 are more recent examples of aircraft using thrust vectoring for control. Figure 2.6 presents the flight control system with conventional control surfaces.

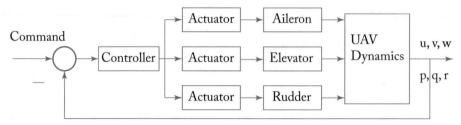

Figure 2.6: Flight control system with conventional control surfaces.

The primary idea behind the design of flight control surfaces is to position them so that they function primarily as moment generators. They provide three types of rotational motion (roll, pitch, and yaw). Variations to this classical configuration lead to some variations in the arrangements of these control surfaces.

With the control effectiveness known (or estimated) for the surfaces controlling each of the three axes mentioned above, then the classical three control/three degrees of freedom system can be defined by an algebraic problem with three equations (the commanded moments) and three unknowns (the required control deflections). Assuming that the mathematical system of equations is consistent, a unique control configuration exists for any desired vector of control generated moments. To apply this methodology, the following information regarding the candidate UAV is needed:

1. UAV configuration and the layout of the major components;

2. mass properties: center of gravity travel, weight, and inertia variations;

3. extreme performance objectives: maximum Mach number versus altitude; maximum load factor and maximum and minimum thrust limits;

4. operational and controllability requirements;

5. type of power transmission; and

6. other systems engineering requirements (e.g., budget constraints, maintainability requirements, production requirements, and reliability requirements)

The design criteria of the control surfaces are determined through the definition of system operational requirements, which in turn evolve from the UAV's mission. The control surfaces may impose limits on the automatic flight control system (e.g., saturation). Although the designed control surfaces may satisfy controllability requirements, several system engineering factors must also be taken into account. These factors include structural considerations (fatigue, flutter, and aero-elasticity), manufacturability, the power to deflect the control surfaces, total cost (manufacturing cost and operating cost), control input saturation, maintainability, reliability, and producibility.

Some parameters must be minimized, some must be maximized, while other ones must be evaluated to ensure that they are acceptable. In some cases, the design of the control surfaces may impose slight to considerable changes to the UAV configuration during the conceptual design process.

Integrated flight control refers to the control and optimization techniques to achieve or improve overall system performance by taking into account subsystem interactions, uncertainties, and even failures for all control effectors. The flight control integration task specifically addresses the issues of interacting impacts between flight control systems and other aircraft functional or structural systems. As an example, conventional control surfaces lose control effectiveness at low speeds and also are ineffective in the post-stall flight regime. Thrust vectoring control has been developed to expand the flight envelope of modern military aircraft.

The integration of system engineering principles with the analysis-driven design process demonstrates that a higher level of integrated control can be attained, identifying the requirements/functional/physical interfaces and the complimentary technical interactions which are promoted by this design process. The objective is to assess the influences of the conceptual design choices and modeling uncertainties on the control system configuration and not so much to predict accurately the absolute value of the control torques.

Figure 2.7: MLB Bat 4, a mini-UAV (image courtesy of avia.pro).

One specific area in which the systems engineering approach has been applied to an integrated flight control system design is the design of integrated flight and propulsion control systems.

For example, in hypersonic vehicle design, the coupling between the propulsion system's dynamics and the aerodynamics is significant enough that an approach designing the propulsion control system and the flight control system separately is not adequate. The strong relationship between the analysis and the influencing parameters allow definite, traceable relationships to be constructed. In the case of control surface design, the major parameters are drawn almost completely from the operational and controllability requirements.

2.8   QUESTIONS

1. What design disciplines will work in parallel within a UAV design project?

2. During the aerodynamic design process, what parameters must be determined?

3. What are the criteria for selection of an airfoil for a wing?

4. What is the primary function of a fuselage?

5. What are the main structural members of a fuselage?

6. What are the main structural members of a wing?

7. What are the three fundamental designs in the wing construction?

8. What is the primary function of an autopilot?

9. In designing the high lift device for a wing, what parameters must be determined?

10. What are the MTOW, and engine power, and wing area Boeing Insitu ScanEagle?

11. In the concept of stealth, what are the three basic methods of minimizing the reflection of pulses back to a receptor?

12. What is the acoustic (i.e., noise) wavelength (signature) range for detecting an air vehicle?

13. Provide at least three main differences between RQ-4A and RQ-4B.

14. What are the two issues for fuel when an aircraft is flying at high altitude?

15. What landing gear parameters need to be determined in the design process?

16. What are the typical engine types for a UAV propulsion system?

17. What are the typical five major stresses to which structural members are subjected?

18. What is the typical propeller efficiency (η_P)?

19. What are the nine configurations for a landing gear?

20. Name typical mechanical/power transmission systems options.

21. What are the conventional control surfaces?

22. Name four unconventional control surfaces.

23. What do FEM and CFD stand for, respectively?

24. What is Nacelle?

25. What are the primary control surfaces of Global Hawk and Predator?

Part II

CHAPTER   3

Fundamentals of Autopilot

3.1   INTRODUCTION

The most important subsystem in a UAV compared with a manned aircraft is the autopilot, since there is no human in a RPV/UAV. Fundamentals are reviewed. The autopilot is a vital and required subsystem within unmanned aerial vehicles. Since there is no human pilot involved in a UAV, an autopilot is a device that must be capable of accomplishing all types of controlling functions including automatic take-off, flying toward the target destination, perform mission operations (e.g., surveillance) and automatic landing. The autopilot has the responsibility to: (1) stabilize the UAV, (2) track commands, (3) guide the UAV, and (4) navigate. This chapter will introduce primary sections of an autopilot, and will review some fundamental topics such as dynamic modeling, UAV dynamics, forces and moments, transfer function, state space, and linearization.

AFCS is responsible for considerable amount of failures/mishaps. The functions of the control and stability of a UAV will depend in nature on the different UAV configurations and the characteristics required of them. A typical UAV Autopilot Design Objective is as follows: Design and evaluate, a robust nonlinear control system that permits the UAV to autonomously fly a complete mission. This control system must account for uncertainties in the UAV model, disturbances in the atmosphere, and measurement noise. A UAV designer must be familiar with basic fundamentals such as trim, control, and stability.

3.2   PRIMARY SUBSYSTEMS OF AN AUTOPILOT

The primary functions of an autopilot are to: (1) stabilize under-damped or unstable modes, (2) accurately track commands generated by the guidance system, (3) guide the UAV to follow the trajectory, and (4) determine UAV coordinates (i.e., navigation). Therefore, an autopilot consists of the: (1) command subsystem, (2) control subsystem, (3) guidance subsystem, and (4) navigation subsystem. These four subsystems must be designed simultaneously to satisfy the UAV design requirements.

In a conventional autopilot, three laws are governing simultaneously in three subsystems: (1) control system through a control law, (2) guidance system via a guidance law, and (3) navigation system through a navigation law. In the design of an autopilot, all three laws need to be designed. The design of the control law is at the heart of autopilot design process. The relation between the control system, the guidance system, and the navigation system is shown in Figure 3.1.

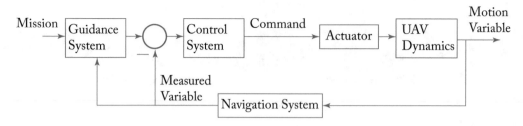

Figure 3.1: Control, guidance, and navigation systems in an autopilot.

3.3   DYNAMIC MODELING

The design of autopilot requires a variety of technical information including the UAV dynamics. The first step to analysis and design of a control system is to describe its dynamic behavior with a mathematical language. The quantitative mathematical description of a physical systems is known as dynamic modeling. There are several ways for mathematical descriptions. The most widely used method is the differential equation.

Description of the behavior and components of a dynamic system with a mathematical language is referred to as the dynamic modeling. Once a physical system has been described by a set of mathematical equations, they are manipulated to achieve an appropriate mathematical format. There are two main techniques used to model a dynamic system—the transfer function and the state space representation. The first one is described in s-domain (frequency); the second one is presented in time domain. A dynamic system may also by represented pictorially; this method is using a block diagram. A UAV is a dynamic system and its dynamic behavior will be modeled by either way. Using a UAV dynamic model, one is able to design automatic flight control system (AFCS) to satisfy the design requirements.

The essential feature of an automatic control system is the existence of a feedback loop to give good performance. This is a closed loop system; if the measured output is not compared with the input the loop is open. Usually it is required to apply a specific input to a system and for some other part of the system to respond in the desired way. The error between the actual response and the ideal response is detected and fed back to the input to modify it so that the error is reduced. The simplest linear closed loop system incorporates a negative feedback, and will have one input and one output variable. Figure 3.2 illustrates a SISO[3] closed-loop system, where K, G(s), and H(s) represent controller, Plant, and measurement device. In general both input and output vary with time, and the control system can be mechanical, pneumatic, hydraulic and electrical in operation, or any combination of these or other power sources.

---

[3]   Single-Input-Single-Output

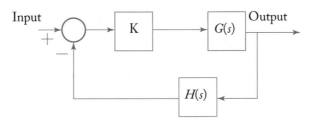

Figure 3.2: A SISO closed-loop system.

Various effective AFCSs have been designed for UAVs using traditional design methods. The synthesis of any AFCS is often costly and requires time-consuming endeavor. This chapter presents the dynamic modeling of the UAV, and enables a UAV designer to use the mathematical tools in the design process.

The control system design process requires the UAV mathematical model as the basis for the design. The UAV modeling process basically consists of: (1) dynamics modeling, (2) aerodynamics modeling, (3) engine modeling, and (4) structural modeling. A UAV is basically a nonlinear system. Furthermore, its dynamics and equations of motion are also nonlinear. The UAV dynamic model is could be linearized or decoupled. The nonlinear coupled equations of motion are the most complete dynamic model. The flat-Earth vector equations of motion are often used, and when they are expanded the standard 6-DOF equations used for UAV control design and flight simulation are obtained. Reference [50] presents the set of nonlinear coupled equations of motion for a UAV. Other UAV dynamic models are: (1) linear decoupled equations of motion, (2) hybrid coordinates, (3) linear coupled equations of motion, and (4) nonlinear decoupled equations of motion.

The aerodynamic forces and moments (aerodynamics model) of the complete UAV are defined in terms of dynamic pressure, aircraft geometry, and dimensionless aerodynamic coefficients. The aerodynamic coefficients are assumed to be the linear functions of state variables and control inputs. These forces and moments are used in equations of motion (dynamic model) as part of control system design, as well as the flight simulation.

The propulsion model is based on the propulsion system powering the UAV (e.g., prop-driven or jet engine). In case of a jet engine, the engine thrust (T) is modeled in terms of throttle setting (δT). In case of a prop-driven engine, the engine power is a function of required engine thrust, aircraft speed, and propeller efficiency. Several control system design techniques may be found in the literature. The choice of the type of control system depends on a variety of factors including the system's characteristics and the design requirements. The heart of the control system is the controller. A summary of the different systems and the different control system design techniques is shown in Figure 4.3 (in Chapter 4). There are several textbooks and papers regarding controller design in the literature; Anderson and Moore [14] and Phillips [15] introduce several controller design techniques.

The real UAV behavior is nonlinear and has uncertainties. It is also important to note that all measurement devices (including gyros and accelerometers) have some kind of noise that must be filtered. It is well known that the atmosphere is a dynamic system that produces lots of disturbances throughout the aircraft's flight. Finally, since fuel is expensive and limited, and actuators have dynamic limitations, optimization is necessary in control system design. Therefore, it turns out that only a few design techniques, such as robust nonlinear control, are able to satisfy all safety, cost, and performance requirements. In order to select the best controller technique, one must utilize a trade-off study and compare the advantages and disadvantage of all candidate controllers.

Virtually all dynamic systems are nonlinear; yet an overwhelming majority of operational control laws have been designed as if their dynamic systems were linear and time-invariant. As long as the quantitative differences in response are minimal (or at least acceptable in some practical sense), the linear time invariant model facilitates the control system design process. This is because of the direct manner in which response attributes can be associated with model parameters.

A small error in modeling, a small error in the control system design, or a small error in the simulation may each result in problems in flight, in the worst case might even result in the loss of an unmanned aerial vehicle. As long as the dynamic effects of parameter variations are slow in comparison to state variation, control design can be based on an ensemble of time-invariant dynamic models. Fast parameters may be indistinguishable from state components, in which case the parameters should be included in an augmented state vector for estimation. The UAV field is one where extensive use is made of modeling and simulation technologies.

The numerical simulation of the aircraft's dynamics is the most important tool in the development and verification of the flight control laws for an aircraft. The availability of special-purpose simulation languages, massive computing capabilities at decreased cost, and advances in simulation methodologies have made simulation one of the most widely used and accepted tools in flight operations research and aircraft systems analysis.

The complete aircraft systems and dynamics model incorporates different subsystem models (e.g., aerodynamics, structures, propulsion, and control subsystems) that have interdependent responses to any input. These subsystems also interact with the other subsystems. The dynamic modeling of an aircraft is at the heart of its simulation. The response of an aerial vehicle system to any input, including commands or disturbances (e.g., wind gusts), can be modeled by a system of ordinary differential equations (i.e., the equations of motion). Dealing with the nonlinear, fully coupled differential equations of motion is not an easy task.

The key component in a low-cost simulation software package is the aircraft dynamics model represented as a set of ordinary differential equations. The dynamics of an aircraft can be modeled in different ways. The equations of motion take several forms (Figure 3.3) including: (1) nonlinear fully coupled, (2) nonlinear semi-coupled, (3) nonlinear decoupled, (4) nonlinear reformulated, (5) linear coupled, (6) linear decoupled, and (7) linear time-variant.

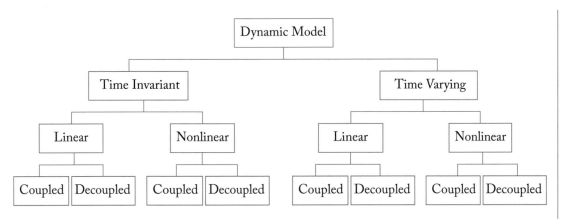

Figure 3.3: Classes of dynamic models.

To develop a computer simulation to evaluate the performance of an aerial vehicle (manned or unmanned) including its control system, we have to invariably use a nonlinear fully coupled model. In order to design a control system, one of the above models are utilized. Each of these models has advantages and disadvantages. These include precision, accuracy, complexity, and credibility. The use of flight simulation tools to reduce risk and flight testing for an aerial vehicle system reduces the overall program schedule.

3.4   UAV DYNAMICS

The UAV dynamics is the model of dynamics of a UAV. The dynamic behavior of a flight vehicle is based on the Newton's second law: When a force is applied on an object, the acceleration of the object is directly proportional to the magnitude of the force, in the same direction as the force, and inversely proportional to the mass of the object. The standard body axis nonlinear fully coupled equations of motion include the three force and the three moment first-order differential equations as follows [2]:

$$\dot{U} = RV - WQ - g \sin \theta + \frac{1}{m} [-D + T \cos \alpha] \tag{3.1a}$$

$$\dot{V} = -UR + WP + g \sin \phi \cos \theta + \frac{1}{m} [Y + T \cos \alpha \sin \beta] \tag{3.1b}$$

$$\dot{W} = +UQ - VP + g \cos \phi \cos \theta + \frac{1}{m} [-L - T \sin \alpha] \tag{3.1c}$$

$$\dot{P} = (c_1 R + c_2 P)Q + c_3 (L_A + L_T) + c_4 (N_A + N_T) \tag{3.1d}$$

$$\dot{Q} = c_5 PR + c_6 (P^2 - R^2) + c_7 M \tag{3.1e}$$

$$\dot{R} = (c_8 P - c_2 R)Q + c_4 (L_A + L_T) + c_9 (N_A + N_T) \tag{3.1f}$$

In the above equations, the c_1 are functions of the moments of inertia, and can be calculated based on the equations of Roskam [1]. The parameters U, V, W are the linear velocity components, and P, Q, R are the corresponding angular rates. The aerodynamic forces D, Y, and L, are the drag, side-force, and lift, and L_A, M_A, N_A, are the aerodynamic moments. The variables α, β, ϕ, and θ are angle of attack, sideslip angle, bank angle, and pitch angle respectively. The motions of a UAV are longitudinal, lateral, and directional. When decoupling is applied to Equation (3.1), the equations of motion are still nonlinear, but decoupled as in the following sections.

1. Longitudinal

Assuming a constant mass, no roll rate, a zero roll angle, no sideslip, and no yaw rate, the nonlinear longitudinal equations of motion reduce to two forces equations and one moment equation:

$$\dot{U} = - WQ - g \sin \theta + \frac{1}{m} [- D + T \cos \alpha] \tag{3.2a}$$

$$\dot{W} = UQ + g \cos \theta + \frac{1}{m} [- L - T \sin \alpha] \tag{3.2b}$$

$$\dot{Q} = \frac{M}{I_Y} \tag{3.2c}$$

2. Lateral-Directional

Assuming zero pitch rate, constant pitch attitude, constant air speed, and constant altitude, the nonlinear lateral-directional equations of motion reduce to one force equation and two moment equations:

$$\dot{V} = g \sin \phi \cos \theta_1 + W_1 P - U_1 R + \frac{1}{m} [Y + T \cos \alpha_1 \sin \beta] \tag{3.3a}$$

$$\dot{P} = c_3 (L_A + L_T) + c_4 (N_A + N_T) \tag{3.3b}$$

$$\dot{R} = c_4 (L_A + L_T) + c_9 (N_A + N_T), \tag{3.3c}$$

where c_3, c_4, and c_9 are introduced in Roskam [1].

3.5   AERODYNAMIC FORCES AND MOMENTS

When the air is passing around an object (e.g., a UAV), the aerodynamic forces and moments are generated. The aerodynamic forces D, Y, and L, are the drag, sideforce, and lift, and L_A, M_A, N_A, are the aerodynamic moments (see Figure 3.4). They are functions of airspeed, air density, wing area, and UAV configuration:

$$D = \bar{q} S C_D \tag{3.4}$$

$$Y = \bar{q} S C_Y \tag{3.5}$$

$$L = \bar{q} S C_L \tag{3.6}$$

$$L_A = \bar{q} S C_l b \tag{3.7}$$

$$M_A = \bar{q} S C_m C \tag{3.8}$$

$$N_A = \bar{q} S C_n b \tag{3.9}$$

where S is the wing reference area, C is the wing mean aerodynamic chord, b is the wing span, and $\bar{q}$ is the dynamic pressure. With the force and moment expressed as linearized functions of the state and control, the nonlinear inverse can be constructed based on these linearized functions and the nonlinear kinematics. The parameters, C_D, C_Y, C_L, C_l, C_m, and C_n are drag coefficient, side-force coefficient, lift coefficient, rolling moment coefficient, pitching moment and yawing moment coefficients respectively.

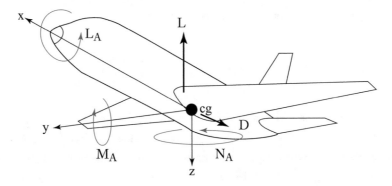

Figure 3.4: Aircraft axes and aerodynamic axes, forces, and moments.

3.6   STABILITY AND CONTROL DERIVATIVES

The drag, side-force, lift, rolling moment, pitching moment, and yawing moment coefficients are all functions of motion variables, UAV weigh and geometry, and configuration. The aerodynamic coefficients are modeled with respect a new group variable referred to as the stability and control derivatives:

$$C_D = C_{D_0} + C_{D_\alpha}\,\alpha + C_{D_q}Q\frac{C}{2U_1} + C_{D_{\dot\alpha}}\,\dot\alpha\,\frac{C}{2U_1} + C_{D_u}\frac{u}{U_1} + C_{D_{\delta_e}}\,\delta_e \tag{3.10a}$$

$$C_y = C_{y_\beta}\beta + C_{y_p}\,P\frac{b}{2U_1} + C_{y_r}\,R\frac{b}{2U_1} + C_{y_{\delta_a}}\,\delta_a + C_{y_{\delta_r}}\,\delta_r \tag{3.10b}$$

$$C_L = C_{L_0} + C_{L_\alpha}\,\alpha + C_{L_q}Q\frac{C}{2U_1} + C_{L_{\dot\alpha}}\,\dot\alpha\,\frac{C}{2U_1} + C_{L_u}\frac{u}{U_1} + C_{L_{\delta_e}}\,\delta_e \tag{3.10c}$$

$$C_l = C_{l_\beta}\,\beta + C_{l_p}\,P\,\frac{b}{2U_1} + C_{l_r}\,R\,\frac{b}{2U_1} + C_{l_{\delta_a}}\,\delta_a + C_{l_{\delta_r}}\,\delta_r \tag{3.10d}$$

$$C_m = C_{m_0} + C_{m_\alpha}\,\alpha + C_{m_q}Q\,\frac{C}{2U_1} + C_{m_{\dot\alpha}}\,\dot\alpha\,\frac{C}{2U_1} + C_{m_u} + \frac{u}{U_1} + C_{m_{\delta_e}}\,\delta_e \tag{3.10e}$$

$$C_n = C_{n_\beta}\beta + C_{n_p}\,P\frac{b}{2U_1} + C_{n_r}\,R\frac{b}{2U_1} + C_{n_{\delta_a}}\,\delta_a + C_{n_{\delta_r}}\,\delta_r \tag{3.10f}$$

In the Equations (3.10a)–(3.10f), parameters such as $C_{L\alpha}$, C_{Du}, $C_{n\beta}$, C_{lp}, are referred to as stability derivatives. In addition, parameters such as $C_{m_{\delta_e}}$, $C_{n_{\delta_a}}$, and $C_{l_{\delta_r}}$ are referred to as control derivatives. Each derivatives is the partial derivative of an aerodynamic force or moment coefficient with respect to a motion variable or control variable. For instance:

$$C_{L_q} = \frac{\partial C_L}{\partial q} \tag{3.11}$$

$$C_{y_\beta} = \frac{\partial C_y}{\partial \beta} \tag{3.12}$$

A group of the variables which are widely used in the design of control surfaces are control derivatives. The control derivatives are simply the rate of change of aerodynamic forces and moments (or their coefficients) with respect to a control surface deflection (e.g., elevator). Control derivatives represent how much change in an aerodynamic force or moment acting on an aircraft when there is a small change in the deflection of a control surface. The greater a control derivative, the more powerful is the corresponding control surface. Three most important non-dimensional control derivatives are $Cl_{\delta A}$, $Cm_{\delta E}$, $Cn_{\delta R}$. The unit of all non-dimensional control derivatives is 1/ rad. The derivative $Cl_{\delta A}$ is the rate of change of rolling moment coefficient with respect to a unit change in the aileron deflection (Equation (3.13)). The derivative $Cm_{\delta E}$ is the rate of change of pitching moment coefficient with respect to a unit change in the elevator deflection (Equation

(3.14)). The derivative $Cn_{\delta R}$ is the rate of change of yawing moment coefficient with respect to a unit change in the rudder deflection (Equation 3.15).

$$C_{l_{\delta A}} = \frac{\partial C_l}{\partial \delta_A} \tag{3.13}$$

$$C_{m_{\delta E}} = \frac{\partial C_m}{\partial \delta_E} \tag{3.14}$$

$$C_{n_{\delta R}} = \frac{\partial C_n}{\partial \delta_R} \tag{3.15}$$

After the vehicle main components (e.g., wing, tail, and landing gear) are designed, the control power requirements may be expressed and interpreted in terms the control derivatives. For instance a rudder is designed to satisfy the requirements of $C_{n_{\delta R}} < -0.4$ 1/rad for a fighter. Or, an elevator is designed to satisfy the requirements of $C_{m_{\delta E}} < -2$ 1/rad for a transport aircraft. It is a challenging task for a flight dynamic engineer to determine the UAV derivatives accurately. Wind tunnel tests are a beneficial technique to calculate the stability and control derivatives.

3.7 TRANSFER FUNCTION

One format of the mathematical model of a control system is transfer function. The control systems are dynamic in nature, and the mathematical models are usually differential equations. If these equations are linearized, then the Laplace transform can be utilized to develop transfer function and control law. The Laplace transform method is originally used to facilitate and systematize the solution of ordinary constant-coefficient differential equations.

Laplace transform is a tool to convert a time-domain function into a frequency-domain one in which information about frequencies of the function can be captured. It is often much easier to solve problems in frequency-domain with the help of Laplace transform. The Laplace Transform of the function f(t), denoted by F(s) is a function of the complex variable s. The Laplace transform is defined by:

$$\mathcal{L}\left\{f(t)\right\} = F(s) = \int_0^\infty f(t)e^{-st}dt \tag{3.16}$$

If the system differential equation is linear, the ratio of the Laplace transform of output variable to the Laplace transform of input variable, is called the transfer function. In general, the format of a transfer function is the ratio of two polynomials in s:

$$F(s) = \frac{b_1 s^m + b_2 s^{m-1} + \ldots + b_{m+1}}{s^n + a_1 s^{n-1} + \ldots + a_n} \tag{3.17}$$

Moreover, when the roots of numerator (i.e., zeros; z), and the roots of denominators (i.e., poles; p) are determined, the transfer function $F(s)$ could be characterized by its poles and zeros:

$$F(s) = K \frac{(s - z_1) \ldots (s - z_i) \ldots (s - z_m)}{(s - p_1) \ldots (s - p_i) \ldots (s - p_m)} \tag{3.18}$$

For more details, you may refer to math textbooks.

3.8   STATE-SPACE MODEL

In general, state space is defined as the n-dimensional space in which the components of the state vector represent its coordinate axes. A mathematical model in time domain based on matrix operation is referred to as state space model. Basic matrix properties are used to introduce the concept of state and the method of writing and solving the state equations. The state of a system is a mathematical structure containing a set of n variables $x_1(t)$, $x_2(t)$, ..., $x_n(t)$, called the state variables, m inputs, $u_1(t)$, $u_2(t)$, ..,$u_m(t)$, and p outputs, $y_1(t)$, $y_2(t)$, ..., $y_p(t)$. In general, a state space model includes a set of n linear first-order differential equation, plus p linear algebraic equation:

$$\dot{x} = Ax + Bu \tag{3.19}$$
$$y = Cx + Du$$

where $A_{n \times n}$, $B_{n \times m}$, $C_{p \times n}$, and $D_{p \times m}$ are matrices. The state equations of a system are a set of n first-order differential equations, where n is the number of independent states. The UAV dynamics may be represented by state space model. In this case, the input variables are such variables as the elevator, aileron, and rudder deflections. In addition, the output variables are usually motion variables such as linear speeds (u, v, w), angular velocities (p, q, r), and angles such as angle of attack (α), and pitch angle (θ).

3.9   LINEARIZATION

In reality, the UAV dynamics is nonlinear. However, dealing with a nonlinear system is not as easy task. One way to remove the nonlinearity using a linearization technique. The linearization converts a nonlinear model to a linear one. Any nonlinear equation is linearized at only an equilibrium (i.e., trim) point. Equilibrium point is also called trim point. The linearization is valid only at the vicinity of the equilibrium point. The trim point is a point or a motion condition that there is no acceleration (i.e., where the velocity is constant). Aircraft example: cruise with constant speed, climb with constant speed, turn with constant speed. Take-off and landing do not have the trim conditions. A linear system satisfies the properties of superposition and homogeneity. The general form of a linear equation is $y = m\,x$, where m is constant numbers.

A linear term is one which is first degree in the dependent variables and their derivatives. A linear equation is an equation consisting of a sum of linear terms. The linearization about the equilibrium point can be accomplished two ways: (1) Expand the non linear equations in a Taylor series and retain only the linear terms; and (2) Directly substitute expand the non linear equations and assume x and u are small and retain only linear terms.

Using Taylor series expansion, for the case of a function with a single independent variable, $F(X)$ can be represented by the following infinite series:

$$F(X) = F(X_1) + \frac{dF}{dX}\bigg|_{x=x_1} (X - X_1) + \frac{d^2F}{dX^2}\bigg|_{x=x_1} \frac{(X - X_1)^2}{2!} + \dots \tag{3.20}$$

Linearization using Taylor series expansion, ignores the higher-order terms. Thus, one will obtain the following linear function:

$$F(X) = \frac{dF}{dX}\bigg|_{x=x_1} X \tag{3.21}$$

As an application, when the nonlinear dynamics of a UAV (Equation (3.1)) is linearized, the following linearized equations are obtained:

$$m\,(\dot{u} - V_1 r - R_1 v + W_1 q + Q_1 w) = -mg\,\cos\,\Theta_1\theta + f_{A_X} + f_{T_X}$$

$$m\,(\dot{v} + U_1 r + R_1 u - W_1 p - P_1 w) = -mg\,(\sin\,\Phi_1\,\sin\,\Theta_1\theta + \cos\,\Phi_1\,\cos\,\Theta_1\,\phi) + f_{A_Y} + f_{T_Y}$$

$$m\,(\dot{w} - U_1 q - Q_1 u + V_1 p + P_1 v) = -mg\,(\cos\,\Phi_1\,\sin\,\Theta_1\theta - \sin\,\Phi_1\,\cos\,\Theta_1\,\phi) + f_{A_Z} + f_{T_Z} \tag{3.22}$$

$$I_{xx}\,\dot{p} - I_{xz}\,\dot{r} - I_{xz}\,(P_1 q + Q_1 p) + (I_{zz} - I_{yy})\,(R_1 q - Q_1 r) = l_A - l_T$$

$$I_{yy}\,\dot{q} + (I_{xx} - I_{zz})\,(P_1 r + R_1 p) + I_{xz}\,(2P_1 p - 2R_1 r) = m_A + m_T$$

$$I_{zz}\,\dot{r} - I_{xz}\,\dot{p} + (I_{yy} - I_{xx})\,(P_1 q + Q_1 p) + I_{xz}\,(Q_1 r + R_1 q) = n_A + n_T \tag{3.23}$$

where subscript 1 refers to the trim point. Next, by using stability and control derivatives, and converting these linear equations of motion into state-space model, the following linearized equations are produced. They then may be reformatted into matrix form as:[4]

$$E\dot{x} = A^* x + B^* u$$

$$y = Cx + Du \tag{3.24}$$

where

$$A^* = \begin{bmatrix} X_u & X_\alpha & -g\cos\gamma & 0 & 0 & 0 & 0 & 0 \\ Z_u & Z_\alpha & -g\sin\gamma & Z_q + V_T & 0 & 0 & 0 & 0 \\ 0 & 0 & 0 & 1 & 0 & 0 & 0 & 0 \\ M_u & M_\alpha & 0 & M_q & 0 & 0 & 0 & 0 \\ 0 & 0 & 0 & 0 & Y_\beta & g\cos\theta_o & Y_p & Y_p - V_T \\ 0 & 0 & 0 & 0 & 0 & 0 & \dfrac{\cos\gamma_o}{\cos\theta_o} & \dfrac{\sin\gamma_o}{\cos\theta_o} \\ 0 & 0 & 0 & 0 & \mu L_\beta + \sigma N_\beta & 0 & \mu L_p + \sigma N_p & \mu L_r + \sigma N_r \\ 0 & 0 & 0 & 0 & \mu N_\beta + \sigma L_\beta & 0 & \mu N_p + \sigma L_p & \mu N_r + \sigma L_r \end{bmatrix} \tag{3.25}$$

---

[4] Using the format from Roskam [1].

$$
B^* = \begin{bmatrix}
X_{\delta_{th}}\cos\alpha & X_{\delta_e} & 0 & 0 \\
X_{\delta_{th}}\sin\alpha & Z_{\delta_e} & 0 & 0 \\
0 & 0 & 0 & 0 \\
M_{\delta_{th}} & M_{\delta_e} & 0 & 0 \\
0 & 0 & Y_{\delta_a} & Y_{\delta_r} \\
0 & 0 & 0 & 0 \\
0 & 0 & \mu L_{\delta_a}+\sigma N_{\delta_a} & \mu L_{\delta_r}+\sigma N_{\delta_r} \\
0 & 0 & \mu N_{\delta_a}+\sigma L_{\delta_a} & \mu N_{\delta_r}+\sigma L_{\delta_r}
\end{bmatrix},
\quad
E = \begin{bmatrix}
1 & 0 & 0 & 0 & 0 & 0 & 0 & 0 \\
0 & V_T-Z_{\dot\alpha} & 0 & 0 & 0 & 0 & 0 & 0 \\
0 & 0 & 1 & 0 & 0 & 0 & 0 & 0 \\
0 & -M_{\dot\alpha} & 0 & 1 & 0 & 0 & 0 & 0 \\
0 & 0 & 0 & 0 & V_T & 0 & 0 & 0 \\
0 & 0 & 0 & 0 & 0 & 1 & 0 & 0 \\
0 & 0 & 0 & 0 & 0 & 0 & 1 & 0 \\
0 & 0 & 0 & 0 & 0 & 0 & 0 & 1
\end{bmatrix}
$$

The states and controls are thus: $x = [V_T, \alpha, \theta, Q, \beta, \phi, P, R]^T$ and $u = [\delta_T, \delta_E, \delta_A, \delta_R]^T$.

When decoupling and linearization techniques are applied simultaneously to Equation (3.18), the state-space equations are split into two groups, each having four states, two inputs, and four outputs. An important point when applying this form is that they are reliable only in the vicinity of the trim point. The validity of these equations is conversely related to the distance from the trim point. As the flight condition get further from trim point, the validity of the result is reduced.

3.10　AUTOPILOT DESIGN PROCESS

In the preceding sections, primary subsystems of an autopilot are introduced. Moreover, dynamic modeling using two techniques (transfer function and state-space model) including aerodynamic forces and moments have been discussed. In this section, the design process of an autopilot is briefly presented.

Figure 3.5 presents the general design process of an autopilot. The design has an iterative nature and begins with the design requirements. Three major sections of control system, guidance system, and navigation system are designed in parallel. At any stage of the design procedure, the output is checked with the design requirements, and a feedback is taken. The process is repeated until the requirements are met. The primary requirements for the design of an autopilot are as follows: (1) manufacturing technology, (2) required accuracy, (3) stability requirements, (4) structural stiffness, (5) load factor, (6) flying quality requirements, (7) maneuverability, (8) reliability, (9) cost, (10) UAV configuration, (11) maintainability, (12) weight, (13) communication system, (14) aerodynamic considerations, (15) processor, and (16) complexity of trajectory.

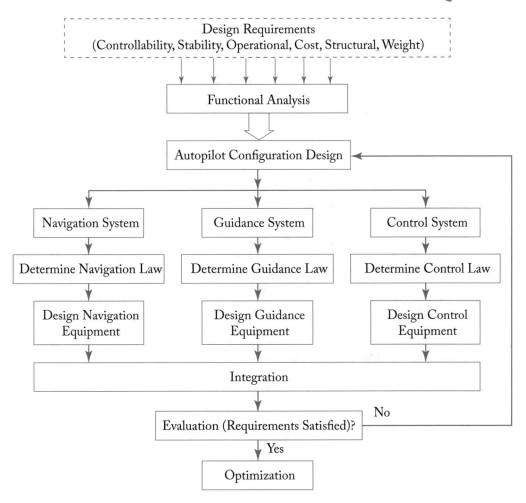

Figure 3.5: Autopilot design process

3.11   QUESTIONS

1. Name primary subsystems of an autopilot.

2. What is the function of the control system?

3. What is the primary function of the guidance system?

4. What is the primary function of the navigation system?

5. Define Laplace transform.

6. Define transfer function.

7. What are the poles and zeros of a transfer function?

8. What is dynamic modeling?

9. What is the name of technique for linearization?

10. Describe the state space model.

11. What is the control derivative?

12. What is the stability derivative?

13. What are the main aerodynamic forces?

14. What are the main aerodynamic moments?

15. Describe autopilot design process.

16. What are the primary requirements for the design of an autopilot?

17. What does the UAV model consist of?

CHAPTER 4

Control System Design

4.1 INTRODUCTION

One of the main subsystems within an autopilot is the control system. The control system controls the direction of the motion of the vehicle or simply the orientation of the velocity vector. The control system is used to keep a UAV on a predetermined course or heading, necessary for the mission. Despite poor and gusty weather conditions, the UAV must maintain a specified heading and altitude in order to reach its destination safely. In addition, in spite of rough air, the trip must be made as smooth as possible for a low load factor. The problem is considerably complicated by the fact that the UAV has six degrees of freedom. This fact makes control more difficult than the control of a ship, whose motion is limited to the surface of the water.

A successful control designer needs not only a good understanding of aerodynamics and flight dynamics, but also a good understanding of the systems engineering approach. All UAVs must meet controllability requirements to be certified for commercial use or adopted by the military. Many military UAVs such as UCAVs also have additional maneuverability requirements. A UAV's ability to meet these requirements is often limited by the amount of control authority available. Thus, it is essential for designers to evaluate the control authority of candidate concepts early in the conceptual design phase.

In general, the primary criteria for the design of control system are as follows: (1) manufacturing technology, (2) required accuracy, (3) stability requirements, (4) structural stiffness, (5) load factor, (6) flying quality requirements, (7) maneuverability, (8) reliability, (9) life-cycle cost, (10) UAV configuration, (11) stealth requirements, (12) maintainability, (13) communication system, (14) aerodynamic considerations, (15) processor, (16) complexity of trajectory, (17) compatibility with guidance system, (18) compatibility with navigation system, and (19) weight.

An essential activity in the detail design is the development of a functional description of the control system components to serve as a basis for identification of the resource necessary for the system to accomplish its mission. Such function may ultimately be accomplished through the use of equipment, software, facilities, data, or various combinations thereof. The control system design must meet the controllability requirements prescribed for maneuvers as desired by the customer or standards. The UAV standards have not been yet finalized, but military specifications (MIL-STD-1797 [5] or UAV certification guidelines), or FAR Part 23 [3] or FAR Part 25 [4] may be referenced.

4.2   FUNDAMENTALS OF CONTROL SYSTEMS

The UAV control system must fundamentally be of a closed-loop form, which employs a negative feedback. A closed-loop system is generally made up of four basic elements:

1. plant,

2. controller,

3. actuator (or servo), and

4. measurement device.

The block diagram of a control system with negative feedback can often be simplified to the form shown in Figures 4.1 and 4.2 where the standard (see Dorf and Bishop [11] and Ogata [12]) symbols and definitions used in feedback systems are indicated. In this feedback control system the output, Y is the controlled variable. This output is measured by a feedback element (measurement device, H) to produce the primary feedback signal, which is then compared with the reference input. The difference between the reference input (R) and the feedback signal (i.e., error), is the input to the controller, K. Then, the controller is generating a signal to the actuator based on a control law.

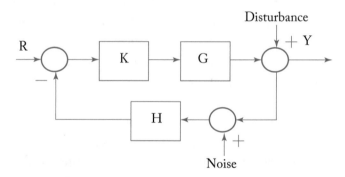

Figure 4.1: Block diagram of a control system including disturbance and noise.

The controller signal is applied to the plant, G via an actuator (e.g., a mechanical jack); actuating signal. The error signal is defined as the ideal or desired system value minus the actual system output. The ideal value establishes the desired performance of the plant. Disturbance is the unwanted signal (e.g., gust) that tends to affect the controlled variable. The disturbance may be introduced into the system at many places. Noise is the unwanted signal (e.g., engine vibration) that tends to affect the measured variable. A filter may be utilized to get rid of noise in the measurement device.

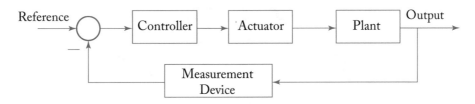

Figure 4.2: Block diagram of a closed-loop control system

In terms of the UAV application, a flight controller in a feedback control system is used to control the aircraft motion. Two typical signals to the system are the reference flight path, which is set by the autopilot, and the level position (attitude) of the airplane. The ultimately controlled variable is the actual course and position of the UAV. The output of the control system, the controlled variable, is the aircraft heading.

The controller is designed based on a control law. Some typical control laws are: (1) linear, (2) nonlinear, (3) optimal, (4) adaptive, and (5) robust. A summary of the dynamic systems; and the different control system design techniques is shown in Figure 4.3. In implementing control law, there are two main approaches: (1) analog control and (2) digital control.

The real aircraft behavior is nonlinear and has uncertainties. A UAV is basically a nonlinear system. Furthermore, its dynamics and equations of motion are also nonlinear. In this section, some common nonlinear systems phenomena are presented. Later on, they are applied to the design of the UAV autopilot. Systems containing at least one nonlinear component are called nonlinear systems. Basically, there are two types of nonlinearities: (1) continuous and (2) discontinuous (hard). Hard nonlinearities include: (1) coulomb friction, (2) saturation, (3) dead-zone, (4) backlash, and (5) hysteresis. In another classification, nonlinearities could be inherent (natural) or intentional (artificial).

It is also important to note that all measurement devices (including gyros and accelerometers) have some kinds of noise that must be filtered. It is well known that the atmosphere is a dynamic system that produces lots of disturbances throughout the aircraft's flight. Finally, since fuel is expensive and limited, and actuators have dynamic limitations, optimization is necessary in control system design. Therefore, it turns out that only a few design techniques, such as robust nonlinear control, are able to satisfy all safety, cost, and performance requirements. However, due to cost, and complexity, many UAV designers adopt a more conventional control architecture. Two conventional controller design tools/techniques are: (1) the root locus technique and (2) frequency domain techniques. The interested reader may refer to Dorf and Bishop [11] and Ogata [12] for more details.

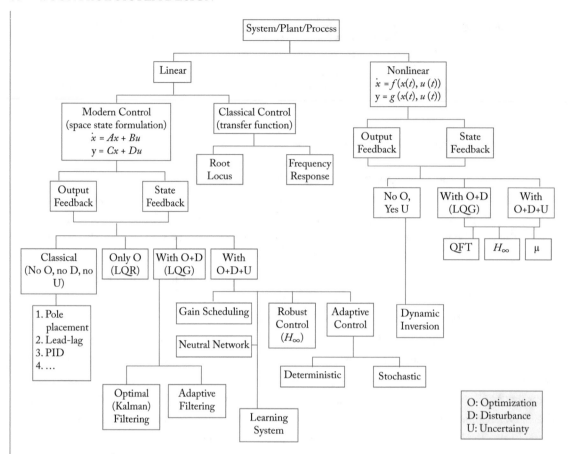

Figure 4.3: Control system design methods.

A regular aircraft actuator for control surfaces is modeled with a first-order transfer function, G_{c}, which is of the form:

$$G_c(s) = \frac{K}{s + K} \tag{4.1}$$

where K represents the inverse of the time constant (τ) of the actuator.

$$\tau = \frac{1}{K} \tag{4.2}$$

A typical value for the time constant of a UAV control surface actuator is about 0.05–0.1 sec, so the K frequently varies between 10 and 20. Recall that the time constant is defined as the time that takes the response of an element to reach 63% of the steady-state value. The smaller the time constant, the faster (more desired) is the actuator.

4.3   UAV CONTROL ARCHITECTURE

Two primary prerequisites for a safe flight are stability and controllability. The control system not only is able to control the UAV, but also is sometimes expected to provide/augment stability. Flight stability is defined as the inherent tendency of an aircraft to oppose any input and return to original trim condition if disturbed. When the summation of all forces along each three axes, and summation of all moments about each three axes are zero, an aircraft is said to be in trim or equilibrium. In this case, aircraft will have a constant linear speed and/or a constant angular speed. Control is the process to change the aircraft flight condition from an initial trim point to a final or new trim point. This is performed mainly by autopilot through moving the control surfaces/engine throttle. The desired change is basically expressed with a reference to the time that takes to move from initial trim point to the final trim point (e.g., pitch rate, q and roll rate, p).

4.3.1   CONTROL CATEGORIES

An aircraft is capable of performing various maneuvers and motions; they may be broadly classified into three main groups: (1) longitudinal control, (2) lateral control, and (3) directional control. In the majority of aircraft, longitudinal control does not influence the lateral and directional control. However, lateral and directional control are often coupled; any lateral motion will often induce a directional motion; and; any directional motion will often induce a lateral motion. The definition of these motions is as follows.

1. **Longitudinal control:** Any rotational motion control in the x-z plane is called longitudinal control (e.g., pitch about y-axis, plunging, climbing, cruising, pulling up, and descending). Any change in lift, drag, and pitching moment have the major influence on this motion. The pitch control is assumed as a longitudinal control.

2. **Lateral control:** The rotational motion control about x-axis is called lateral control (e.g., roll about x-axis). Any change lift distribution and rolling moment have the major influence on this motion. The rolling control is assumed as a lateral control.

3. **Directional control:** The rotational motion control about z-axis and any motion along y-axis is called directional control (e.g., yaw about z-axis, side-slipping, and skidding). Any change in side-force and yawing moment have the major influence on this control. The yaw control is assumed as a directional control. A level turn is a combination of lateral and directional motions.

In a conventional UAV, three primary control surfaces are used to control the physical three-dimensional attitude of the UAV, the elevators, rudder, and ailerons. In a conventional UAV, the longitudinal control (in the x-z plane) is performed through a longitudinal control surface or elevator. The directional control (in the x-y plane) is performed through a directional control surface or rudder. The lateral motion (rolling motion) is executed using aileron. Hence, there is a direct relationship between the control system and control surfaces. A graphical illustration of these three control motions is sketched in Figure 4.4. A UAV has six degrees of freedom (i.e., three linear motions along x, y, and z; and three angular motions about x, y, and z). Thus, there are normally six outputs in an actual flight; six examples are three linear velocities (u, v, w), and three angular rates (p, q, r).

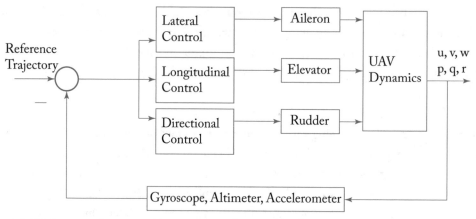

Figure 4.4: Flight control system with conventional control surfaces.

The primary idea behind the design of flight control surfaces is to position the surfaces; so that they function primarily as moment generators. They provide three types of rotational motion (roll, pitch, and yaw). Variations to this classical configuration lead to some variations in the arrangements of these control surfaces. Table 2.2 represents several control surfaces configurations.

There are various measurement devices to measure the flight variables such as airspeed, pitch angle, heading angle, bank angle, linear accelerations (normal, lateral, and longitudinal), angular rates (pitch, roll, and yaw rates), altitude, and position. The measured flight data are recorded, and may be stored in a data storage element; which may be conveniently read by the user (in real-time, or off-line). If an interface with telemetry data systems is available, a user in the ground station may access the data in real time. Typical measurement devices (sensors) are: (1) gyroscope, (2) rate gyroscope, (3) pitot-tube, (4) altimeter, (5) magnetometer, (6) compass, (7) accelerometer, and (8) GPS.

The directional gyroscope is used as the error-measuring device. Two gyros must be used to provide control of both heading and attitude (level position) of the airplane. The error that appears in the gyro as an angular displacement between the rotor and case is translated into a voltage by various methods, including the use of transducers such as potentiometers. Augmented stability for the UAV may be desired in the control system by rate feedback. In other words, in addition to the primary feedback, which is the position of the airplane, another signal proportional to the angular rate of rotation of the airplane around the vertical axis is fed back in order to achieve a stable response. A "rate gyro" is used to supply this signal.

4.3.2   CRUISE CONTROL

The UAV cruise control is assume to be the easiest control, while has several alternatives. In the lift equation (3.6), the independent parameters that are: UAV weight (W), airspeed (V), altitude or its corresponding air density (ρ), and angle of attack, or its associated lift coefficient (C_L). Since the fuel is consumed during flight, the aircraft weight is constantly decreased during the flight. In order to maintain a level flight, we have to decrease the lift as well. Of the many possible solutions only three are more practical and will be examined. In each case, two flight parameters will be held constant throughout cruise. The three options of interest for continuous decrease of the lift during cruise are:

1. decreasing flight speed (constant-altitude, constant-lift coefficient flight);

2. increasing altitude (constant-airspeed, constant-lift coefficient flight); and

3. decreasing angle of attack (constant-altitude, constant-airspeed flight).

For each flight program, a separate controller is designed. In the first option, the velocity must be reduced with the same rate as the aircraft weight is decreased. In the second solution, the air density must be decreased; in another word, the flight altitude must be increased. The third option offers the reduction of aircraft angle of attack; i.e., the reduction of lift coefficient. In terms of autopilot operation, the first option is applied through throttle; and the third option is implemented through stick/yoke/wheel. In the second option, no action is needed by the pilot; the aircraft will gradually gain height (climbs).

Based on the safety regulations and practical considerations, the second option is option of interest for majority of aircraft. In general, when flight is conducted under the jurisdiction of Federal Aviation Regulations, the accepted flight program is the constant-altitude, constant-airspeed flight program.

In the first option, for the view-point of autopilot control, there are three drawbacks to this flight program. The first is the need to continuously compute the airspeed along the flight path and to reduce the throttle setting accordingly. The second is that reducing the airspeed increases the flight times. The third is the fact that air traffic control rules require "constant" true airspeed for cruise flight, currently constant means ± 10 knots. The good news is that, current autopilot of large transport aircraft has solved part of this problem (i.e., no need for autopilot calculation).

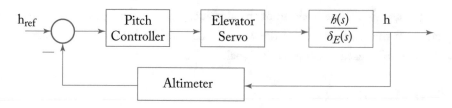

Figure 4.5: Block diagram of altitude control system.

The second flight program is commonly referred to as *cruise-climb* flight. In this option, the air density will be automatically reduced as the aircraft weight is decreased. No autopilot intervention is necessary. Therefore, cruise-climb flight requires no computations or efforts by the pilot. In an aircraft equipped with an autopilot, the aircraft cruise control will be implemented by the autopilot. After establishing the desired cruise airspeed, the pilot simply engages the Mach-hold mode (or constant-airspeed mode) on the autopilot; and the aircraft will slowly climb at the desired flight-path angle as the fuel is burned.

Figure 4.5 illustrates the block diagram of altitude control system. For maintaining the constant cruise altitude, either of elevator or throttle may suffices. One option is to keep the throttle contestant (engine thrust). Thus, the elevator is employed to change the angle of attack for varying UAV weight. The pitch controller could be as simple as a PID or to use a more complicated one. The measurement device is an altimeter (either thorough GPS, or pitot-tube, or radar altimeter). In either case, the UAV will maintain the constant altitude through a longitudinal control system.

4.4   FLIGHT CONTROL REQUIREMENTS

There are primarily three flight control requirements: (1) longitudinal control requirements; (2) lateral/roll control requirements, and (3) directional control requirements. In this section, these requirements are presented. These requirements must be met by the flight control system. The description and analysis of aircraft modes show that Automatic Flight Control Systems (AFCSs) can be divided into different categories. One category includes modes that involve mainly the rotational degrees of freedom, whereas some categories involve the translational degrees of freedom.

4.4.1   LONGITUDINAL CONTROL REQUIREMENTS

An aircraft must be longitudinally controllable, as well as maneuverable within the flight envelope (Figure 4.6). In a conventional aircraft, the longitudinal control is primarily applied through the deflection of elevator (δ_E) and engine throttle setting (δ_T). There are two groups of requirements in the aircraft longitudinal controllability: the required stick force and the aircraft response to the pilot input. In order to deflect the elevator, the pilot must apply a force to stick/yoke/wheel and hold it (in the case of an aircraft with a stick-fixed control system). In an aircraft with a stick-free control system, the pilot force is amplified through such devices as tab and spring. The stick force analysis is out of scope of this text; the interested reader is referred to study references such as Roskam [1].

The aircraft response in the longitudinal control is frequently expressed in terms of pitch rate (q). However, the forward speed and angle of attack would be varied as well. The most critical flight condition for pitch control is when the aircraft is flying at a low speed. Two flight operations which feature a very low speed are take-off and landing. Take-off control is much harder than the landing control due to the safety considerations. A take-off operation is usually divided into three sections: (1) ground section, (2) rotation or transition, and (3) climb. The longitudinal control in a take-off is mainly applied during the rotation section which the nose is pitched up by rotating the aircraft about main gear.

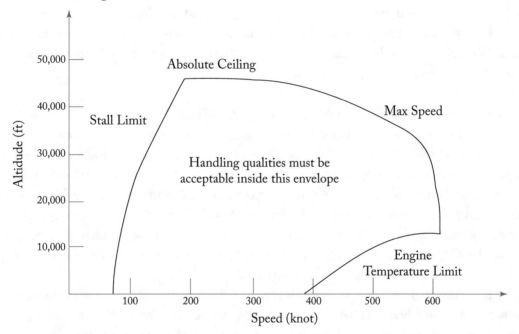

Figure 4.6: A typical operational flight envelope.

The control surfaces must be designed such that aircraft possesses acceptable flying qualities anywhere inside the operational flight envelope (for example); and allowable cg range, and allowable aircraft weight. The operational flight envelopes define the boundaries in terms of speed, altitude and load factor within which the aircraft must be capable of operating in order to accomplish the desired mission. A typical operational flight envelope for a transport aircraft is shown in Figure 4.6.

4.4.2   ROLL CONTROL REQUIREMENTS

Roll or lateral control requirements govern the aircraft response to the aileron deflection; thus, the requirements are employed in the design of aileron. It is customary to specify roll power in terms of the change of bank angle achieved in a given time in response to a step function in roll command. Thus, the aircraft must exhibit a minimum bank angle within a certain specified time in response to aileron deflection. The required bank angles and time are specified in Tables for various aircraft classes and different flight phase (see MIL-F-8785C [6]).

Roll performance in terms of a bank angle change ($\Delta\phi$) in a given time (t) is specified in Tables (MIL-F-8785C [6]) for Class I–IV aircraft. The notation "60° in 1.3 sec" indicates the maximum time it should take from an initial bank angle (say 0°) to reach a bank angle which is 60° different than the initial one, following the full deflection of aileron. It may also be interpreted as the maximum time it should take from a bank angle of -30° to +30°. For Class IV aircraft, for Level 1, the yaw control should be free. For other aircraft and levels, it is permissible to use the yaw control to reduce any sideslip which tends to retard roll rate. Such yaw control is not permitted to induce sideslip which enhances the roll rate.

4.4.3   DIRECTIONAL CONTROL REQUIREMENTS

In a conventional aircraft, directional control is usually maintained by the use of aerodynamic controls (e.g., rudder) alone at all airspeeds. There are a number of cases that directional control must be achievable within a specified limits and constraints. In this section, most important ones are presented. Directional control characteristics shall enable the pilot to balance yawing moments and control yaw and sideslip. Sensitivity to yaw control pedal forces shall be sufficiently high that directional control and force requirements can be met and satisfactory coordination can be achieved without unduly high pedal forces, yet sufficiently low that occasional improperly coordinated control inputs will not seriously degrade the flying qualities.

In a multi-engine aircraft, at all speeds above 1.4 Vs with asymmetric loss of thrust from the most critical factor while the other engine(s) develop normal rated thrust, the airplane with yaw control pedals free may be balanced directionally in steady straight flight. The trim settings shall be those required for wings-level straight flight prior to the failure. When an aircraft in is directional trim with symmetric power/thrust, the trim change of propeller-driven airplanes with speed shall be such that wings-level straight flight can me maintained over a speed range of ±30% of the trim speed or ±100 knots equivalent airspeed, whichever is less (except where limited by boundaries of the Service Flight Envelope) with yaw-control-device (i.e., rudder). In the case of one-engine-in-operative (asymmetric thrust), it shall be possible to maintain a straight flight path throughout the Operational Flight Envelope with yaw-control-device (e.g., rudder) not greater than 100 pounds for Levels 1 and 2 and not greater than 180 pounds for Level 3, without re-trimming.

4.5   PID CONTROLLER

One form of controller widely used in industrial process control called a three-term, or PID controller. In this controller, three operations are applied on the error signal: (1) proportionally (P) amplified, (2) integrated (I), and (3) differentiated (D). Thus, the control signal, u(t), in time domain is:

$$u(t) = K_p \, (e(t)) + K_I \int e(t)dt + K_D \frac{de(t)}{dt} \tag{4.3}$$

Hence, the controller has three terms: (1) proportional, (2) integral, and (3) derivative. In s-domain, this controller has a transfer function:

$$G_c(s) = K_p + \frac{K_I}{s} + K_D s \tag{4.4}$$

Various performance deficiencies may be corrected by employing the right values of PID gains. This type of controller is effective, low cost, and easy to apply. Thus, it is even used in aircraft autopilot. Dorf and Bishop [11] and Ogata [12] provide the technique to determine PID gains.

4.6   OPTIMAL CONTROL–LINEAR QUADRATIC REGULATOR (LQR)

The optimal control [14] is based on the optimization of some specific performance criterion. In this technique, no disturbance, noise, or uncertainty is considered. The performance of a control system, written in terms of the state variables, can be expressed as:

$$J = \int_0^{t_f} g(x, u, t)dt \tag{4.5}$$

We are interested in minimizing the error of the system; any deviation from equilibrium point is considered an error. To this end, an error-squared performance index is defined. For a system with one state variable, x_1, we have

$$J = \int_0^{t_f} [x_1(t)]^2 dt \tag{4.6}$$

An optimization technique for a dynamic system in state-space format is defined. The Linear Quadratic Regulator (LQR) is an optimal controller. The LQR problem is simply defined as follows. The system of interest is of the form:

$$\dot{x} = Ax + Bu$$
$$, x(0) = x_o. \tag{4.7}$$
$$y = Cx + Du$$

Given the matrices Q and R, the design task is to find the optimal control signal u(t) such that the quadratic cost function:

$$J = \int (x^T Q x + u^T R u) dt \tag{4.8}$$

is minimized. The solution to this problem is: $u = -Kx$, where $K = R^{-1}B^T P$, and P is the unique, positive semi-definite solution to the Algebraic Riccati Equation (ARE):

$$PA + A^T P + Q - PBR^{-1}B^T P = 0. \tag{4.9}$$

Based on this technique, the LQR gains are calculated using a MATLAB program, and then a control system is designed. The flight simulation may be executed by a SIMULINK model to analyze the response. The matrices Q and R are the weight for state and input variables, respectively. They are determined based on the cost function.

4.7   ROBUST CONTROL

The linear robust (H_∞) control technique can be applied to any linear system, either Jacobian or feedback linearized. In this approach, disturbances, noise, and uncertainty; ΔG (see Figure 4.7) are considered. Furthermore, an optimization technique is employed to minimize the infinity norm of the error transfer function. Consider a system described by the state-space equations:

$$\dot{x} = Ax + B_1 w + B_2 u$$
$$z = C_1 x + D_{12} u \tag{4.10}$$
$$y = C_2 x + D_{21} w$$

The desire is to design the feedback control $u = K(s)\, y$, such that $\|T_{zw}(s)\|_\infty < \gamma$ for a given positive number γ. Note that γ is a function of the maximum singular value of the unstructured uncertainty (in fact., $\bar{\sigma}[\|\Delta G\|_\infty] = \frac{1}{\gamma}$). The controller (solution) is given (Doyle and Glover [13]) by the transfer function:

$$K(s) = -F(sI - \hat{A})^{-1} ZL \tag{4.11}$$

where

$$\hat{A} = A + \frac{1}{\gamma^2} B_1 B^T X + B_2 F + ZLC_2 \tag{4.12}$$

and

$$F = -B^T X, L = -YC^T, Z = \left(I - \frac{1}{\gamma^2} YX \right)^{-1} \tag{4.13}$$

where X and Y are solutions of pairs of Algebraic Riccati equations (AREs). The closed loop transfer function matrix $T_{zw}(s)$ from the disturbance w to the output z is given by:

$$T_{zw}(s) = G_{11} + G_{12} K (I - G_{22} K)^{-1} G_{21}, \tag{4.14}$$

where

$$G(s) = \begin{bmatrix} 0 & D_{12} \\ D_{21} & 0 \end{bmatrix} + \begin{bmatrix} C_1 \\ C_2 \end{bmatrix} (sI - A)^{-1} (B_1, B_2) = \begin{bmatrix} G_{11} & G_{12} \\ G_{21} & G_{22} \end{bmatrix} \qquad (4.15)$$

A robust controller can handle uncertainty, disturbance, and noise.

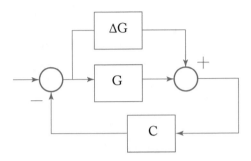

Figure 4.7: Closed-loop system with additive and multiplicative perturbations.

4.8 DIGITAL CONTROL

In the early history of automatic flight control systems, all aspects of the flight control were analog, including the controller. With microprocessors so fast, flexible, light, and inexpensive, the control laws could be implemented in digital form. With the introduction of computers and microprocessors in the 1970s, the modern aircraft take advantage of digital control. Digital control is a branch of control theory that employs computers/microcontrollers for acting as controllers. In the digital control, a computer is responsible for the analysis and implementation of the control algorithm. A digital control system may also use a microcontroller to an application-specific integrated circuit. Since a digital device only accepts digital signal, a sampler (a kind of switch) is required to take the samples of the continuous signal. The samples are in the forms of zero (0) and one (1).

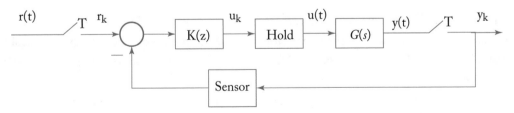

Figure 4.8: Digital control system.

Usually, a digital control system consists of three main elements: (1) an A/D conversion for converting analog input to digital format for the machine, (2) D/A conversion for converting digital output to a form that can be the input for a plant, and (3) a digital controller in the form of a computer, microcontroller or a programmable logic controller.

The digital controller in implemented using a software code in a computer. The schematic of a digital control system is shown in Figure 4.8, where z is the Z-transform variable.

The hold device is a D/A converter the discrete control samples ($K(z)$) into the continuous-time control. The sampler with sample period T is an A/D converter that takes the samples y_k = y (kT) of the output of G(s). In the digital control, the transfer functions are in z-domain (i.e., discrete). In a discrete (digital) system, the Laplace transform is replaced with the z-transform. The relationship between variable s and variable z is:

$$z = e^{sT} \tag{4.16}$$

where T is the sample rate (e.g., 0.01 sec). The approximation of the exponential function is:

$$e^{sT} \approx \frac{1 + sT/2}{1 - sT/2} \tag{4.17}$$

This is referred to as bilinear transformation or Tustin's approximation. Inverting this transformation yields:

$$s = \frac{2}{T} \frac{z - 1}{z + 1} \tag{4.18}$$

In the digital control, an approximate discrete equivalent of every transfer function (plant, sensor, and controller) are obtained by this transformation technique. As the sampling rate (the average number of samples, T) is higher, the approximation gets more accurate. Phillips et al. [15] present the analysis and design of digital control systems. Software packages such as matlab[5] is recommended for the simulation of digital control system. The features and applications of the microcontroller are presented in Chapter 7.

4.9   STABILITY AUGMENTATION

The fundamental function of a UAV control system is to control the flight variables to make sure the UAV is flying in the predetermined trajectory. In a UAV, since there is no human pilot involved, an autopilot is a device that must be capable of accomplishing all types of controlling functions including automatic take-off, flying toward the target destination, and automatic landing. However, for a lightly stable UAV, the secondary function is to provide augmented stability. As the name implies, a stability augmentation system is to augment the stability of an open-loop plant.

The widening performance envelope (e.g., Figure 4.6) created a need to augment the stability of the UAV's dynamics over some parts of the flight envelope. Because of the large changes in the UAV's dynamics, a dynamic model that is stable and adequately damped in one flight condition may become unstable, or at least inadequately damped in another flight condition. For such cases, an autopilot is as an automatic control system that also provides stability augmentation.

Some autopilots improve the stability of the aircraft (SAS), while some augment the response to a control input (CAS). The slow modes (e.g., phugoid and spiral) are controllable by a human pilot. But since it is undesirable for a pilot to have to pay continuous attention to controlling these modes, an automatic control system is needed to provide "pilot relief." Table 4.1 lists typical common types of autopilots. Phillips et al. [15] presents automatic flight control system design and provides several detailed examples.

In this section, one stability augmentation application of an autopilot; yaw damper, is presented. The yaw damper is to augment the directional stability of a UAV by damping the yawing oscillation, known as Dutch-roll. The purpose of the yaw-damper feedback is to use the rudder to generate a yawing moment that opposes any yaw rate that builds up from the dutch roll mode. Most large UAVs with high-altitude cruising flight are equipped with such mode. The reason is that at high altitude, the turbulence are strong, and gusts hits the fuselage nose, and pushes the nose to the left and right almost continuously. The autopilot mode will keep the heading in the desired direction.

Figure 4.9 depicts the block diagram of a yaw damper. The reference yaw rate is desired to be zero ($\dot{\psi} = 0$), so the yaw damper will drive any undesired yaw rate to zero. A rate gyro will measure the yaw rate, and provide a feedback for the system. The controller will produce a controlling signal which is implemented through an actuator (i.e., rudder servo). The yaw damper controller can be as simple as a gain, K. When the UAV tends to turn, and has a bank angle (ϕ_1), the yaw damper will try to fight with the constant bank angle turn. This is due to the fact that the computed yaw rate is a function of bank angle:

$$R_1 = \dot{\psi}_1 \cos(\theta_1)\ \cos(\phi_1) \tag{4.19}$$

where subscript 1 stands for steady-state value. One solution for such case is to use a washout filter with a time constant of about 4 sec. This will create a lag in the response of the rate gyro.

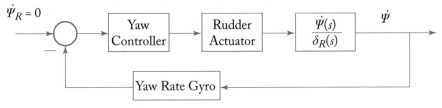

Figure 4.9: Block diagram of a yaw damper.

On May 25, 2017, the General Atomics Aeronautical Systems SkyGuardian MQ-9B (a Certifiable version of its Predator B; Figure 4.10) had an endurance of 48.2 hr. The autopilot had been configured in an Intelligence, Surveillance, and Reconnaissance (ISR) for the entire mission.

Figure 4.10: General Atomics MQ-9 Reaper (sometimes called Predator B).

Table 4.1: Autopilot categories. States (Variables) to be controlled (α, β, θ, γ, φ, ψ, P, Q, R, U, V, W, x, y, h, M, n_x, n_y, n_z)

1	2		3		4
	Hold Functions		**Navigation Functions**		
Stability Augmentation Systems (SAS)[1]	**Longitudinal**	**Lateral-Directional**	**Longitudinal**	**Lateral-Directional**	**Command Augmentation Systems (CAS)**
Roll Damper	Pitch attitude hold (θ)	Bank angle hold (ϕ); wing leveler	Automatic flare mode	Localizer	Command tracking[2] 1. Pitch rate CAS 2. Roll rate CAS
Yaw Damper	Altitude hold (h)	Heading angle hold (ψ)	Glide slope hold	VOR hold	Command generator tracker[3] (model following)
Pitch Damper	Control wheel steering mode	Turn rate mode at constant speed and altitude	Approach categories and guidance	Turn coordination (zero lateral acceleration)	Normal acceleration CAS (n_z)
			Automatic landing		

[1] SAS can be used simultaneously with manual control (pilor command). They improve stability and stabilize an unstable aircraft. [2] Following a non-zero reference command signal. [3] Time-varying trajectories.

4.10   AUTONOMY

A typical full flight consists of the following phases: (1) ground taxi, including ground collision avoidance; (2) take-off, (3) climb; (4) en-route cruise; (5) turn and maneuver, (6) descent, and (7) landing; (8) ground operation at the destination; and (9) handling of emergencies in any of these sectors. Autonomy is defined with respect to every flight phase, and with various levels. Moreover, some high-level autonomy such as detect-and-avoid, fault monitoring, and automated recovery will be addressed in this section.

4.10.1   CLASSIFICATION

As the name implies, there is no human located in an unmanned vehicle. In general, there are four ways for piloting a UAV: (1) remote control, (2) autopilot-assisted control (i.e., automated), (3) semi-full autonomy, and (4) full autonomy. Under full autonomous control, the reality is that the on-board computer is in control not a human operator. A minimal autopilot system includes attitude sensors and onboard processor.

Due to the nonlinearities and uncertainties of the aerial vehicle dynamics, a lot of advanced control techniques, such as neural network, fuzzy logic, sliding mode control, robust control, and learning systems have been used in autopilot systems to guarantee a smooth desirable flight mission. Nowadays, technological advances in wireless communication and micro electromechanical systems, make it possible to use inexpensive small autopilots. We need to first distinguish between these three modes. Here a definition for each mode is provided.

1. In the remote control mode, a human operator is controlling the UAV from a ground station. He/she is making decision, applying input, and providing commands.

2. In the automated or automatic system, in response to feedbacks from one or more sensors, the UAV is programmed to logically follow a pre-defined set of rules in order to provide an output. Knowing the set of rules under which it is operating means that all possible output are predictable. A modern autopilot, allows the vehicle to fly on a programmed flight paths without human interference for almost all the mission, without an operator doing anything other than monitoring its operation.

3. Currently, there is no consensus for the definition of autonomy in UAV community. In this book, "autonomy" is the ability of an agent to carry out a mission in an independent fashion without requiring human intervention. Autonomy should include kind of artificial intelligence, since the decision-making is performed by the autopilot. An autonomous vehicle is capable of understanding higher-level intents and directions. Such a vehicle is able to take appropriate action to bring about a desired state/trajectory, based on this understanding and its perception of the environment,

It is capable of deciding a course of action, from a number of alternatives, without depending on human oversight and control, although these may still be present (for monitoring). Although the overall activity of an autonomous UAV is predictable, individual actions may not be. An autonomous UAV is able to monitor and assess its status (e.g., altitude, airspeed) and configuration (e.g., flap deflection); and command and control assets onboard the vehicle.

The core components of autonomy are command, control, navigation, and guidance. Higher levels of autonomy, which reduces operator workload, include (in increasing order) sense-and-avoid, fault monitoring, intelligent flight planning, and reconfiguration. An autonomous behavior includes observe, orient, decide, and act. The aviation industry objective is that eventually autonomous UAV will be able to operate without human intervention across all flight sectors. Such objective requires advances in various technologies including guidance system, navigation system, control system, sensors, avionics, communication systems, infrastructures, and software microprocessors. Table 4.2 illustrates reliability and mishap rates for several manned and unmanned aircraft. Current accident rate for a typical UAVs is about 50 times greater than that of a General Dynamics fighter F-16 Fighting Falcon. Autopilot is responsible for considerable amount of failures/mishaps. In the autonomy, operator tells the system what is desired from the mission (not how to do it) with the flexibility of dynamic changes to the mission goals being possible in flight with minimal operation re-planning.

Table 4.2: Reliability and mishap rates for several manned/unmanned aircraft

No	Aircraft	Mishap rate (per 100,000 hr)	Reliability (%)
1	General Aviation (e.g., Cessna 172)	1.22	N/A
2	AV-8B Harrier II	10.7	N/A
3	Lockheed U-2 Dragon Lady	3	96.1
4	General Dynamics F-16 Fighting Falcon	3.5	96.6
5	McDonnell Douglas F/A-18 Hornet	3.2	N/A
6	Boeing 747	0.013	98.7
7	Boeing 777	0.013	99.2
8	General Atomics RQ - 1 Predator	32	89
9	Northrop Grumman RQ - 4 Global Hawk	160	N/A

4.10.2 DETECT (I.E., SENSE)-AND-AVOID

Collision avoidance is a primary concern to the FAA regarding aircraft safety. UAVs are seen as potential key airspace users in the future of air transportation, which necessitates additional research and study of safety measures. One of the major limitations to the widespread use of unmanned vehicles in civilian airspace has been the detect-and-avoid problem. In manned civilian aviation, "see-and-avoid"[6] is the primary mechanism by which piloted aircraft avoid collisions with each other. Obviously this is impractical for widespread use of unmanned vehicles, so they must achieve an equivalent level of safety/reliability/assurance to that of manned aircraft operations. There is currently a large amount of research projects being conducted in the area of detect-and-avoid. Active solutions include the use of machine vision, and GPS/radar to detect collision threats, and precision control to avoid collision. However, the current adapting technologies are not at such reliable level [52]. In addition, high computational requirement is another obstacle. A major design issue of UAVs is that any black box is additional weight, and weight restrictions for some smaller UAVs may restrict UAV functionality or the inclusion of cooperative systems.

4.10.3 AUTOMATED RECOVERY

Another challenge in the execution of automated control is the automated recovery. Here the recovery could be either regular automated landing or to recover by some means such as a net. Since physical pilot control is not present, there is a high potential for unsuccessful recovery. The UAV must have a number of fail-safes in place in case of any elemental failure. Another source of failure is communication failure or link loss. In the event that command and control links have been completely severed between a UAV and the ground station, the UAV should be switched to pre-programmed mode to attempt for some fixed period of time to re-establish communications, or to independently complete the mission. This is another big area of research for several research institutes and UAV industries.

4.10.4 FAULT MONITORING

Unscheduled UAV maintenance creates a lot issues and cost for large UAV operator units, because spare parts are not always available at any place and sometimes have to be shipped across the world. Moreover, if a flight mission has to be canceled or even delayed, the UAV causes significant costs. Hence, reducing the number of unscheduled maintenance is a great cost factor for UAV operators. For this objective, and to ensure the integrity of the UAV systems, fault monitoring must be continually conducted on flight. Fault monitoring ensures that undetected system faults will not lead to a catastrophic failure of the aircraft's systems which may eventually lead to human casualties on the ground.

   Failure prediction is the combination of condition (i.e., health) monitoring and condition prediction to forecast when a failure will happen. The objectives of fault monitoring are: (1) reduction of unscheduled maintenance, (2) advanced failure prediction, (3) condition monitoring, (4)

---

[6]   Currently, the traffic alert and collision avoidance system (TCAS) is the primary cooperative collision avoidance system and is in use by a variety of airspace users [51].

ability to better plan maintenance, and, finally, (5) prevent failures. In the event of system faults, the UAV must have the capability to reconfigure itself and re-plan its flight path in a fail-safe manner. The predictive UAV health monitoring is able to predict failures so that maintenance can be planned ahead.

4.10.5 INTELLIGENT FLIGHT PLANNING

An intelligent UAV system must have the ability to plan and re-plan its own flight path, to cope with undesired situations. This results in the requirement for advance sensors, and a high level computing environment where flight planning algorithms can be executed. The intelligent flight planning requires a significant improvement in software and hardware performance. The flight planning process requires knowledge of the UAV's surroundings; including airspace, terrain, other traffic, weather, restricted areas obstacles and closest airfield. The UAV must plan the optimal route for its mission, considering the local environment, to minimize the flight time and fuel usage. The intelligent planning will detect any incoming aircraft for collision avoidance.

4.10.6 MANNED-UNMANNED TEAMING

Today's aircraft inventory includes a diverse mix of manned and unmanned systems. The statistics is growing exponentially. Unmanned aircraft systems (UAS) are subject to regulation by the FAA to ensure safety of flight, and safety of people and property on the ground. Incidents involving unauthorized and unsafe use of small, remote-controlled aircraft have risen dramatically. Pilot reports of interactions with suspected unmanned aircraft have increased from 238 sightings in all of 2014 to 780 through August of 2015. One of the main goals for the manned-unmanned teaming is to provide flexible flight operations. Teaming a UAV system with manned systems will offer advantage to both.

To achieve the full potential of unmanned systems at an affordable cost, efforts must be conducted to implement technologies and evolve tactics, techniques and procedures that improve the teaming of unmanned systems with the manned aircraft. The functions of a UAV in a team with manned aircraft depend in nature on the different UAV configurations and their characteristics. To this end, the critical challenges must be identified for further growth to fulfill expanding UAV roles in supporting the aviation safety goals. Moreover, new technologies need to be developed, and new regulations must be prepared.

4.11   CONTROL SYSTEM DESIGN PROCESS

This additional stabilization may be necessary for a lightly stable UAV. A block diagram of the aircraft control system illustrates control of the airplane heading by controlling the rudder position. In this system the heading that is controlled is the direction the airplane would travel in still air. The pilot corrects this heading, depending on the crosswinds, so that the actual course of the airplane coincides with the desired path. Another control included in the complete airplane control system controls the ailerons and elevators to keep the airplane in level flight.

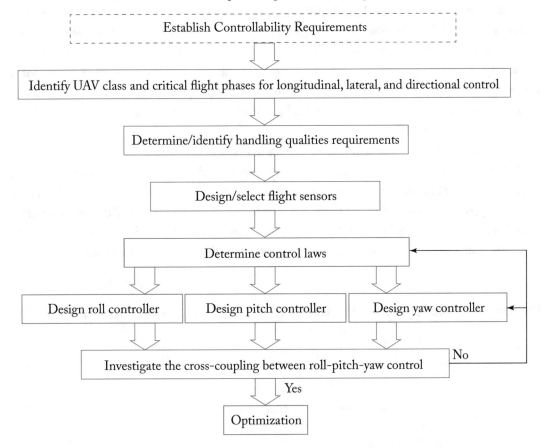

Figure 4.11: Control system design process.

Figure 4.11 illustrates a flowchart that represents the control system design process. In general, the design process begins with a trade-off study to establish a clear line between stability and controllability requirements and ends with optimization. During the trade-off study, two extreme limits of flying qualities are examined and the border line between stability and controllability is drawn. For instance, a fighter UAV can sacrifice the stability to achieve a higher controllability and maneuverability. Then, an automatic flight control system may be employed to augment the aircraft stability. In the case of a civil UAV, the safety is the utmost goal; so the stability is clearly favored over the controllability.

The results of this trade-off study will be primarily applied to establishing the most aft and the most forward allowable location of aircraft center of gravity. Three roll control, pitch control, and yaw control are usually designed in parallel. Then the probable cross coupling between three controls is studied to ensure that each control is not negating controllability features of the aircraft in other areas. If the cross coupling analysis reveals an unsatisfactory effect on any control plane, one or more control systems must be redesigned to resolve the issue. Flight control systems should be designed with sufficient redundancy to achieve two orders of magnitude more reliability than some desired level. In general, the control system performance requirements are: (1) fast response, (2) small overshoot, (3) zero steady-state error, (4) low damping ratio, (5) short rise time, and (6) short settling time. If the overshoot is large, a large load factor will be applied on structure (due to an increase in acceleration).

In FAR 23, one level of redundancy for control system is required (i.e., power transmission line). The lines of power transmission (i.e., wire and pipe) should not be close to each other, should not be close to fuel tanks, and should not be close to hydraulic lines. In most Boeing aircraft, there are three separate hydraulic lines. If there is a leak in the hydraulic lines or if the engines get inoperative, there is an extra hydraulic system which is run independently. So, Boeing 747 has four hydraulic systems. These design considerations provide a highly safe and reliable aircraft.

Example 4.1

Problem statement: A control system is modeled by the following block diagram (Figure 4.12).

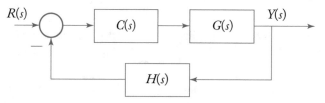

Figure 4.12: The control system of Example 4.1.

where $G(s) = \dfrac{s + 10}{2s^2 + 2s + 9}$; $H(s) = \dfrac{10}{s + 11}$

Design a PID controller ($C(s)$) such that the response to the unit-step reference input satisfies the following performance specifications:

    a. steady state error less than 10%,

    b. maximum overshoot less than 20%,

    c. settling time less than 3 sec, and

    d. rise time less than 1 sec.

Solution:

A Simulink model (by matlab) is created (Figure 4.13). The PID gains are determine to be:
1. $K_p = 5$, $K_i = 0.1$, $K_d = 1$.

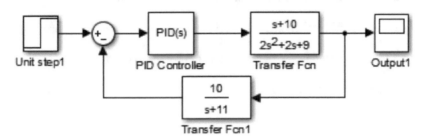

Figure 4.13: Simulink model of the closed-loop system.

The simulation results (Figure 4.14) illustrate that all four design requirements are met.

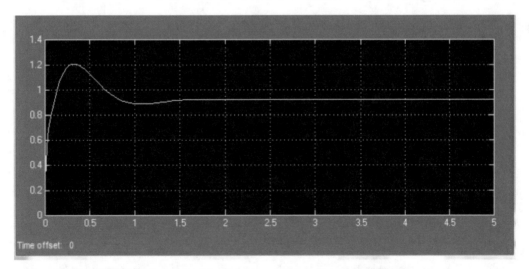

Figure 4.14: The response of the system to a unit step input.

4.12   QUESTIONS

1. What are the primary criteria for the design of control system?

2. Name four control laws.

3. Name four basic elements of a closed-loop system.

4. Name three nonlinearities of a control systems.

5. What are the two conventional controller design tools/techniques?

6. Write the typical mathematical model for a regular actuator for control surfaces.

7. What is a typical value for the time constant of a UAV actuator?

8. Define longitudinal control.

9. Define lateral control.

10. Define directional control.

11. Describe the flight envelope.

12. Name typical UAV measurement devices (sensors).

13. Name three conventional control surfaces.

14. Describe PID controller.

15. Describe optimal control

16. Describe robust control.

17. Describe digital control.

18. What A/D and D/A stand for?

19. List autopilot categories/modes.

20. What is the function of a yaw damper?

21. What are the main issues concerning integration of unmanned aerial vehicles in civil airspace?

4.13   PROBLEMS

Perform the following problems for the Cessna 182 at maximum speed. All necessary data such as moment of inertia, and non-dimensional longitudinal stability and control derivatives are given in pages 480-482 of Roskam [1].

1. Calculate the following dimensional longitudinal stability and control derivatives:

$X_u, Z_u, M_u, X_\alpha, Z_\alpha, M_\alpha, X_{\delta_e}, Z_{\delta_e}, M_{\delta_e}, Z_{\dot\alpha}, M_{\dot\alpha}, Z_q, M_q.$

2. Determine the aircraft longitudinal transfer functions: $\dfrac{u(s)}{\delta_e(s)}, \dfrac{\alpha(s)}{\delta_e(s)}, \dfrac{\theta(s)}{\delta_e(s)},$

3. Is the aircraft longitudinally statically stable? Why?

4. Apply -2° of elevator deflection,

   a. Plot the response (forward speed, angle of attack, and pitch angle) of the open-loop aircraft system to this input (for 200 sec). You may use MATLAB for simulation.

   b. What are the new steady state values for angle of attack, pitch angle, and aircraft speed?

   c. Is this aircraft still cruising or climbing?

   d. Determine the rise time, settling time, and maximum overshoot of each response (α, θ, and u).

5. Calculate the state-space dynamic model of the aircraft.

6. Determine its eigen values. Is the aircraft longitudinally dynamically stable? Why?

7. Apply -1° of elevator deflection.

   a. Plot the response (forward speed, angle of attack, and pitch angle) of the open-loop aircraft system to this input (for 300 sec). You may use MATLAB for simulation.

   b. What are the new steady state values for angle of attack, pitch angle, and aircraft speed?

   c. Is this aircraft still cruising or climbing?

8. Design a closed-loop control system including a controller to follow the following trajectory (Figure 4.15). Provide your Simulink model; and the simulation results including a comparison between the required trajectory and the flown trajectory.

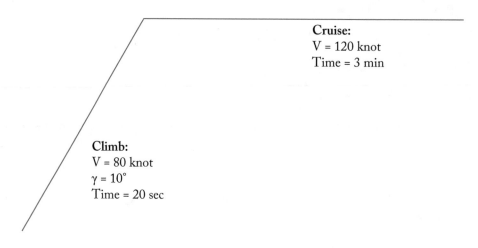

Cruise:
V = 120 knot
Time = 3 min

Climb:
V = 80 knot
$\gamma = 10°$
Time = 20 sec

Figure 4.15: Required trajectory.

9. Calculate the lateral-directional state-space dynamic model of the aircraft.

10. Determine its lateral-directional eigenvalues. Is the aircraft laterally directionally dynamically stable? Why?

11. Apply -1° of aileron/rudder deflections.

   a. Plot the response (bank angle, sideslip angle, heading angle, roll rate, and yaw rate) of the open-loop aircraft system to this input (for 3 sec). You may use MATLAB for simulation.

   b. What are the values for bank angle, sideslip angle, heading angle after 3 sec?

12. For the Cessna 182, design a closed-loop control system including a controller to follow the following trajectory (Figure 4.16). Provide your Simulink model; and the simulation results including a comparison between the required trajectory and the flown trajectory.

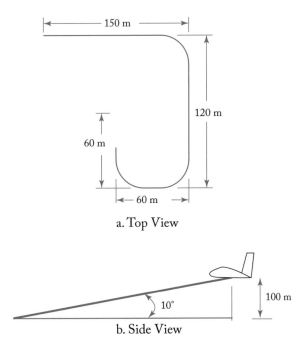

Figure 4.16: The descent and landing trajectory.

13. Determine the aircraft state space model for four inputs, eight state variables, and five outputs. The input variables are u = $[\delta_T, \delta_E, \delta_A, \delta_R]^T$ and the state variables are x = $[V_T, \alpha, \theta, q, \beta, \varphi, p, r]^T$. The output variables are forward speed, angle of attack, pitch angle, sideslip angle, bank angle.

14. Apply simultaneous deflections of -2° of elevator, 11° of aileron, and +2° of rudder (only for 1 sec). Plot the response (forward speed, angle of attack, pitch angle, pitch rate, sideslip angle, bank angle, roll rate, yaw rate and **heading angle**) of the open-loop aircraft system to this input (for 10 sec). You may use MATLAB for simulation.

CHAPTER 5

Navigation System Design

5.1 INTRODUCTION

In general, navigation is the skill that involves the determination of position and direction of a moving objet. More specifically, navigation is a field of study that focuses on the process of monitoring the movement of a vehicle from one place to another. A UAV navigation system is one that determines the position of the air vehicle with respect to some reference frame (Figure 5.1), for example, the Earth sea level. Navigation systems may be entirely onboard a UAV, or they may be located elsewhere and communicate via radio or other signals with the UAV, or they may use a combination of these methods. Navigation is usually done using sensors such as gyros, accelerometers, altimeter and the Global Positioning System (GPS).

Position has mainly three coordinates: (1) latitude, (2) longitude, and (3) altitude above sea level. The latitude of a place on Earth is its angular distance north or south of the equator. Latitude is usually expressed in degrees ranging from 0° at the equator to 90° at the North and South poles. Similarly, the longitude of a place on Earth is the angular distance east or west of the Greenwich meridian. Longitude is ranging from 0° at the Greenwich meridian to 180° east and west. Direction involves two main data: (1) heading and (2) pitch.

Calculation in a navigation system is performed based on a navigation law. Three major classes of navigation laws/systems are: (1) electronic navigation law/system, (2) inertial navigation law/system, and (3) celestial navigation law/system. Electronic navigation system may be performed by three electronic devices: (1) radio, (2) radar, and (3) satellite. Hence, the global positioning system is a type of electronic navigation system. The inertial navigation is classified as a type of dead reckoning, which consists of extrapolation of a known position to some future time. Celestial navigation system is based on observation of the positions of the Sun, Moon, and other stars/plants.

Figure 5.1: Coordinates of a UAV are determined by navigation system at any point.

This chapters presents basic fundamentals, classifications, and features of navigation systems. In addition, two more popular navigation system, the inertial navigation system and GPS, are described with more details. Some related topics such as coordinate systems, avionics, gyroscope, and filtering reviewed in brief. In the end, design considerations for the navigation system is presented.

A standard UAV navigation system often relies on GPS and inertial sensors (INS). If the GPS signal for some reason becomes unavailable or corrupted, the state estimation solution provided by the INS alone drifts in time and will be unusable after a few seconds (especially for small-size UAVs which use low-cost INS). The GPS signal also becomes unreliable when operating close to obstacles due to multi-path reflections. In addition, it is quite vulnerable to jamming (especially for a GPS operating on civilian frequencies). By choosing the right of coordinates system, the navigation equations of motion (Equations 5.1–5.3) are integrated.

5.2   COORDINATE SYSTEMS

Prior to description of the navigation laws and systems, we need to provide the definitions of a few related terms. The motion of a UAV can be described by means of vectors in three dimensions. Coordinate System is a measurement system for locating points in space, set up within a frame of reference. The frame of reference: a rigid body or set of rigidly related points that can be used to establish distances and directions. Inertial frame is defined as a frame of reference in which Newton's laws apply.

In general, there are two groups of frames and coordinate systems: the inertial frame (Earth-fixed), its origin is at the Earth's center of mass (nonrotating but translating with Earth) and the body frame. Three axes of the inertial frame are north, east, and down. The body frame is fixed at the UAV, is moving and rotating with the UAV. Three coordinate systems are classified as body frame: (1) body-fixed system, (2) wind-axes system, and (3) stability-axes system. The orthogonal axes are named x, y, and z. Figure 2.4 shows the axes and positive convention for body frame, and the body-fixed coordinate system. The inertial navigation system is employing the earth-fixed coordinate system; so the position is always defined with respect a fixed point in Earth (e.g., ground system). In the short range UAVs, we may loosely assume the Earth is flat for easy calculations.

5.3   INERTIAL NAVIGATION SYSTEM

Inertial navigation is the oldest navigation technique which is originally developed by naval navigators. This technique is still applicable for the cases where the GPS signals are not available; or sometimes as the parallel (redundant) system for increasing reliability. Inertial navigation is a type of dead reckoning that calculates the position based on onboard motion sensors. Dead reckoning consists of extrapolation of a known position to future period. It involves measurement of time, direction of motion, airspeed, and ground speed. The calculation is performed by taking the last known position and time, considering average speed and heading since then and the present time. The distance travelled is equal to the velocity times the duration of motion. So, the speed is multiplied by the time elapsed since the last position to get distance traveled. This can be added to the initial position to get the present position. The longitude and latitude is determined by using north and east components of the ground speed. Once the initial latitude and longitude is established, the system receives signals from sensor that measure the acceleration along three axes. Hence, the current latitude and longitude are continually calculated.

Inertial navigation was used in a wide range of applications including the navigation of aircraft, missiles, spacecraft, submarines and ships. Inertial navigation systems were in wide use until GPS became available. Inertial navigation system is still in common use on submarines, since GPS reception or other fix sources are not possible while submerged. Its main advantage over other navigation systems is that, once the starting position is set, it is independent, and does not require outside information. In addition, it is not affected by adverse weather conditions and it cannot be detected or jammed. Its main disadvantage is that since the current position is calculated solely from previous positions, its errors or deviations are increasing at a rate proportional to the time elapsed. Inertial navigation systems must therefore be frequently corrected with other devices/techniques.

Inertial navigation is a self-contained navigation technique in which measurements are used to track the position and orientation of an object relative to a known starting point, orientation and velocity. The two main measurement devices for inertial navigation are: (1) accelerometers and (2) gyroscopes. By processing signals from these devices it is possible to calculate the position and orientation of a device. Recent advances in the construction of micro-machined electro-mechanical system devices have made it possible to manufacture small and light inertial navigation systems. For instance, inertial measurement unit (IMU) contains three orthogonal rate-gyroscopes and three orthogonal accelerometers, measuring angular velocity and linear acceleration respectively. To perform these functions the system requires the following instruments: (1) three speed measuring devices (accelerometers, after integration), (2) three angular sensors (gyroscopes), (3) a clock, and (4) a processor.

Inertial measurement units fundamentally fall into two categories: stable platform systems and strap-down systems. In stable platform systems, the inertial sensors are mounted on a platform which is isolated from any external rotational motion, and therefore output quantities measured in the global frame. In strap-down systems the inertial sensors are mounted rigidly onto the UAV, and therefore output quantities measured in the body frame. Stable platform and strap-down systems are both based on the same underlying principles. Strap-down systems have reduced mechanical complexity and tend to be physically smaller than stable platform systems. These benefits are achieved at the cost of increased computational complexity. As the cost of computation has decreased strap-down systems have become the dominant type of inertial measurement units. The acceleration signals in both systems are similar, so when integrated, the same velocity are resulted.

To keep track of velocities, the signals from the accelerometers are integrated. The orientations are determined by the integration of the rate gyro signals. The platform mounted gyroscopes detect any platform rotations. These signals are fed back to torque motors which rotate the gimbals in order to cancel out such rotations, hence keeping the platform aligned with the initial frame. In order to improve overall noise and drift performance, gyros and accelerometers have rather high gains. The mechanism and governing laws of gyroscopes and accelerometers are presented in Section 5.5. For more information on technical details of inertial navigation, study of Grewal et al. [24] is recommended.

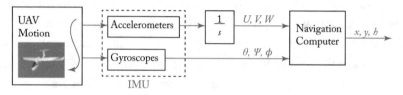

Figure 5.2: Navigation process.

In principle, the calculations of the UAV position is based on the ground speed (V), and the elapsed time (t). The velocity is defined as the time derivatives of the displacement. If the UAV is assumed as a point mass, the new location (x) is simply determined by:

$$x = x_0 + Vt \tag{5.1}$$

However, the velocity of a UAV has three components. Moreover, there are three attitudes in the flight. Thus, the navigation equations [2] which are integrated by the navigation computer are:

$$\dot{x} = U\cos\theta\cos\Psi + V(-\cos\phi\sin\Psi + \sin\phi\sin\theta\cos\Psi) + W(\sin\phi\sin\Psi + \cos\phi\sin\theta\cos\Psi) \tag{5.2}$$

$$\dot{y} = U\cos\theta\sin\Psi + V(\cos\phi\cos\Psi + \sin\phi\sin\theta\sin\Psi) + W(-\sin\phi\cos\Psi + \cos\phi\sin\theta\sin\Psi) \tag{5.3}$$

$$\dot{h} = U\sin\theta - V\sin\phi\cos\theta - \cos\phi\cos\theta \tag{5.4}$$

where U, V, and W stand for velocity in x, y, and z direction; and θ, Ψ, and ϕ denote pitch, heading, and bank angles respectively. The integration delivers x, y, and h which are the UAV position in the global frame. The orientation parameters θ, Ψ, and ϕ are directly measured by gyros. In addition, the velocity parameters U, V, and W are calculated by integration of the signals measured by accelerometers. Figure 5.2 illustrates the navigation process block diagram, and the functions of its two measurement devices.

5.4 GLOBAL POSITIONING SYSTEM

The Global Positioning System (GPS) is a space-based navigation system owned by the United States government and funded, controlled, and operated by the US Department of Defense. The GPS concept is based on time and the known position of originally 24 satellites. There are millions of civil users of GPS worldwide. (including UAVs and RC airplanes). By the introduction of GPS, the altitude measurement has become an easy task and rather more accurate. The GPS has almost no direct effect from atmospheric condition. No matter if an aircraft is flying in an ISA or non-ISA condition, its altitude is measured precisely via GPS. Without radar altimeter or GPS, the measured altitude is subject to an error. This must be taken into account when we are calculating aircraft performance. Many new aircraft are currently equipped with a GPS receiver. So their altitude measurement is accurate, and we do not need to take into account the effect of air density variations.

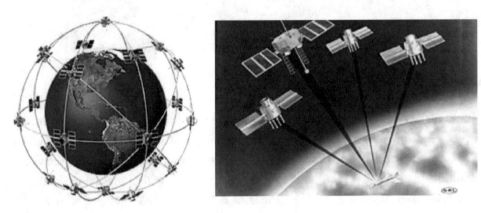

Figure 5.3: GPS satellites (image courtesy of U.S. State Department).

The space segment of the system consists of the GPS satellites. These space vehicles send radio signals from space. The nominal GPS Operational Constellation consists of 24 satellites (Figure 5.3) that orbit the earth in 12 hr [24]. There are often more than 24 operational satellites as new ones are launched to replace older satellites. The satellite orbits repeat almost the same ground track once each day. The orbit altitude (20,200 km, 55° inclination) is such that the satellites repeat the same track and configuration over any point approximately each 24 hr.

There are six orbital planes (with nominally four space vehicles in each), equally spaced (60° apart), and inclined at about 55° with respect to the equatorial plane. This constellation provides the user with between five and eight satellites visible from any point on the earth. The predictable accuracy is within a few meters horizontally and vertically and has a 200 nsec time accuracy.

GPS provides specially coded satellite signals that can be processed in a GPS receiver, enabling the receiver to compute position (e.g., altitude), velocity and time. Four GPS satellite signals are used to compute positions in three dimensions and the time offset in the receiver clock. Position in XYZ is converted within the receiver to geodetic latitude, longitude, and height above the ellipsoid. Velocity is computed from change in position over time, or the Doppler frequencies. The GPS satellites broadcast at the same two frequencies, 1.57542 GHz (L1 signal) and 1.2276 GHz (L2 signal). Because the satellite signals are modulated onto the same L1 carrier frequency, the signals must be separated after demodulation.

Figure 5.4: General Atomics RQ/MQ-1 Predator A.

The availability of GPS has permitted UAV operation to be extended in range, enabling the operation of very long range UAVs. However, the jamming of GPS signals in the event of hostilities is a major concern. In addition, there are times and areas of the earth, that the GPS signal is not available or weak. One solution is to integrate the GPS with a inertial navigation system. For instance, General Atomics RQ/MQ-1 Predator A (Figure 5.4) is equipped with a GPS-aided inertial navigation system.

5.5   POSITION FIXING NAVIGATION

The position fixing navigation is the determination of the position of the UAV (a fix) without reference to any former position. There are three basic methods of fixing position: (1) map reading, (2) celestial navigation, and (3) measuring range and/or bearing to identifiable points.

5.5.1   MAP READING

Map reading involves matching what can be seen of the outside world with a map. It is the traditional method of position fixing on land; and also by general aviation in clear weather. A camera is necessary to capture a picture of outside world. Modern UAVs adopting this technique uses a radar to obtain a picture of the ground from the air and a computer matches it with a map stored in the form of a digital database. These systems are called terrain referenced navigation aids.

5.5.2   CELESTIAL NAVIGATION

Celestial navigation has been used by mariners for centuries. The basic principle of celestial navigation is that if the altitude of a celestial object (measured in terms of the angle between the line-of-sight and the horizontal) of a celestial object is measured; then the observer's position must lie on a specific circle (called the circle of position) on the surface of the earth centered on the point on the earth which is directly below the object. If the time of observation is noted and the celestial object is a star, then this circle can easily be found using astronomical charts and tables. Sightings on two or more such celestial objects will give two or more such circles of position, and their intersection will give the position of the UAV. This technique has been abandoned nowadays in favor of modern navigational aids. This technique may be employed in the areas that GPS signals are not available.

5.6   INERTIAL NAVIGATION SENSORS

The navigation system requires a number of measurement devices (sensors) to measure various flight parameters including velocity, positions, and attitude. When all sensors are located in one set, it is refeed to as inertial measurement unit (IMU). Some important sensors are: (1) accelerometer, (2) gyroscope (or gyro), (3) rate gyro, and (4) magnetometer.

The early models of IMUs were all mechanical, and had a very heavy weight (tens of kg). The new generations are all electronic, and weigh as low as a few grams. A typical current and modern IMU contain one accelerometer, rate gyro, and magnetometer per axis for each of the three vehicle axes. The IMUs are part of avionics that help a UAV to conduct a successful flight. Table 5.1 demonstrates the primary functions of a few navigation sensors. The first two sensors (i.e., accelerometers and gyroscopes) are described with more details in this section.

Table 5.1: Primary functions of a few navigation sensors

No	Name	Measures	Remarks
1	Accelerometer	Linear acceleration	The linear acceleration is converted to linear velocity.
2	Basic gyroscope	Attitude	Based on gyro law
3	Rate gyro	Angular velocity	The angular velocity can be converted to angular positions.
4	Magnetometer	Attitude	e.g., heading
5	Pitot tube	Altitude, airspeed	Using air pressure
6	Compass	Magnetic north	Heading angle is measure w.r.t. north

In general, there are two groups of sensors: (1) sensor as a payload and (2) sensors necessary for navigation/guidance systems. In this chapter, only the sensors which are necessary for a regular flight (such as the ones used in navigation system) have been introduced. Payload sensors are addressed in Chapter 10.

5.6.1   ACCELEROMETER

An accelerometer is a measurement device to measure linear acceleration of a motion. The final flight variable which is derived from the output signal of an accelerometer is the linear velocity. This convention is performed by the process of integration (either by an analog device or by a digital circuit). A secondary integration of such signal produces the position of the flight vehicle (i.e., x, y, and z). Table 5.2 lists accelerometers and their outputs.

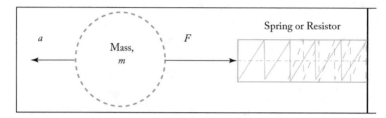

Figure 5.5: Elements of an early basic mechanical accelerometer.

The early version of mechanical accelerometer (Figure 5.5) was purely mechanical and worked based on the Newton's second law. Currently, there are variety of accelerometers employing laser, magnetic field, beam deflection, heat and temperature, optics, frequency, micro-machined piezoelectric, and electric signal. In the purely mechanical accelerometer, the acceleration of the UAV creates an inertial force (F) on the opposite direction to the mass within the acceleration (a). The Newton's second law:

$$F = ma \tag{5.5}$$

Table 5.2: Primary functions of a few navigation sensors

No	Accelerometer	Axis	Acceleration
1	Along x-axis	Longitudinal axis	n_x
2	Along y-axis	Lateral axis	n_y
3	Along z-axis	Normal/vertical axis	n_z

When the moving mass (m) inside the accelerometer is known, and the force (F) is measured, the acceleration is simply the force divided by the mass. The force applied to the mass can also be measured by various elements such as a mechanical spring. In this case, the spring deflection is an indication of the applied force. The spring deflection can also be measured by a change in the resistance of a resistor, which can be represented by a voltage. The presentation of the theory behind non-mechanical accelerometers is out of scope of this text. Today, various analog and digital accelerometers are available in the market that can measure the acceleration of up to 250 g. The maximum acceleration of a UAV is typically less than 20 g. The size of a modern accelerometer is about a few millimeter, and its price tag is a few dollars.

5.6.2 GYROSCOPE

Another important measurement device for a navigation system is gyroscope (or simply gyro). A gyroscope is a mechanical sensor to measure the change in attitude (heading angle, bank angle, and pitch angle) and/or its rate of the UAV. There are two groups (see Table 5.3) of gyros: attitude gyro and rate gyro. As the name implies, an attitude gyros is incorporating an attitude reference (e.g., bank angle), so it is used to sense deviations from a reference attitude in the aircraft y-z, x-y, and y-z planes. However, a rate gyro will measure the rate of change of an attitude (e.g., yaw rate) with respect to a reference plane. The pitch-rate gyroscope measures the (inertial) angular rate around the y axis (q). The roll-rate gyroscope measures the angular rate around the x axis (p). The yaw-rate gyroscope measures the angular rate around the z axis (r).

The output of the rate gyro is a signal proportional to the angular velocity of the case about its input axis with respect to inertial space. The sensitivity of the rate gyro is such that it cannot detect the component of earth's rate parallel to the input axis of the gyro. Thus, the gyro effectively measures the angular velocity of the case, and thus the aircraft, with respect to the earth. The gyro filter is usually necessary to remove noise and/or cancel structural mode vibrations.

Table 5.3: Gyros and their outputs				
No	Gyroscope	Axis	Measure	Reference Plane	Symbol
1	Attitude gyro	About x-axis	Bank angle	y-z plane	ϕ
2		About y-axis	Pitch angle	x-z plane	θ
3		About z-axis	Heading angle	x-y plane	ψ
4	Rate gyro	About x-axis	Roll rate	y-z plane	p
5		About y-axis	Pitch rate	x-z plane	q
6		About z-axis	Yaw rate	x-y plane	r

Gyroscope is a spatial mechanism (as shown in Figure 5.6) and is composed of a spinning rotor, a fixed frame, and two gimbals (inner and outer). The rotor in spinning inside the inner gimbal, which in turn is free to rotate inside the outer gimbal. In Figure 5.6, the rotor is spinning about z-axis. When the plane of the rotor is rotated about one of its axis (say x), the gyroscope is creating a force along the other axis (y). The resultant force will be measures; and is an indication of the change in attitude.

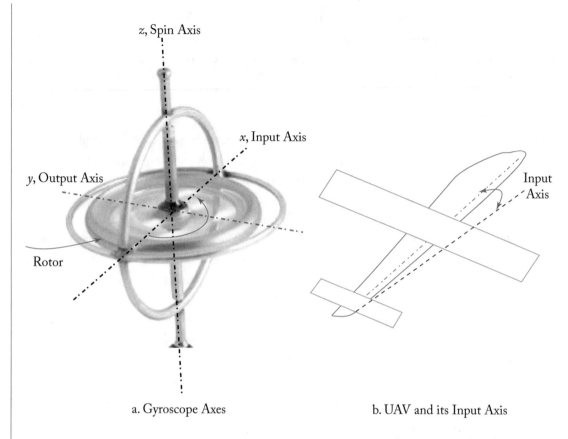

a. Gyroscope Axes   b. UAV and its Input Axis

Figure 5.6: Gyroscope (based on Retro Gyroscope from mastgeneralstore.com).

The gyro is working based on the law of the gyro (gyro law) which can be derived from Newton's law in rotational form. In principle, if the axis of rotation of a gyroscope is displaced, a nutation is produced. The law states that the time rate of change with respect to inertial space of the angular momentum (H) of an object about its center of gravity is equal to the applied torque (i.e., the angular momentum is conserved). This can be written in equation form as

$$M_{app} = \frac{d}{dt} H_I = I \frac{d}{dt} \omega \tag{5.6}$$

where I is the mass moment of inertia of the rotor, and ω is its angular speed. As Equation (5.5) involves the time derivative of a vector with respect to inertial space, the equation of Coriolis must be applied.

$$\frac{d}{dt} H_I = \frac{d}{dt} H_E = \omega_{IE} \times H \tag{5.7}$$

where ω_{IE}, is the angular velocity of the earth with respect to inertial space (0.07292115×10^{-3} rad/sec). The gyroscope itself can be mounted on a base that is moving with respect to the earth. In addition, the case of the gyroscope can be mounted on a platform so that it can rotate relative to the base. Finally the inner gimbal can rotate relative to the case. Applying governing equations of these rotations, and inserting the relevant terms into Equations 5.5 and 5.6 yield the "law of the gyro."

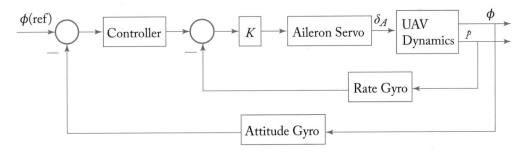

Figure 5.7: A roll-angle-hold autopilot.

In modern UAVS, a rate gyro is employed in conjunction with an attitude gyro for each axis. For instance, in many cases, a roll-rate feedback is used, in conjunction with the roll-angle feedback to control the bank angle. Figure 5.7 demonstrates the block diagram of a roll-angle-hold autopilot which illustrates the control system along with the navigation system. In the navigation system, a rate gyro is utilized in the inner loop, while an attitude gyro is used in the outer loop. Having two feedbacks (roll rate, p; and bank angle, ϕ) will provide a precision control. An alternative element to attitude gyro is the magnetometer, which is sensitive to magnetic fields.

5.6.3   AIRSPEED SENSOR

The GPS is a powerful tool in providing UAV ground speed, but it cannot measure the airspeed. The reason is that the atmosphere is a dynamic system, and often gust and wind are present. For safety reasons, the knowledge of airspeed is always crucial and necessary. A simple tool to measure airspeed is a standard pitot-static tube provided that it is suitably positioned to read accurate static pressure either as part of a combined unit ahead of any aerodynamic interference or as a separate static vent elsewhere on the aircraft. The errors involve in pressure reading include position error, installation error, and calibration issues.

There is an inaccuracy in the classic pitot-static tube in measuring airspeed, as well as the inability to record speeds below about 15 m/sec. In addition, the fluctuating reading values from pitot tube can cause instability in the control system. This, it is recommended to rely on data from a system integrated with GPS that does not require knowledge of ambient static pressure.

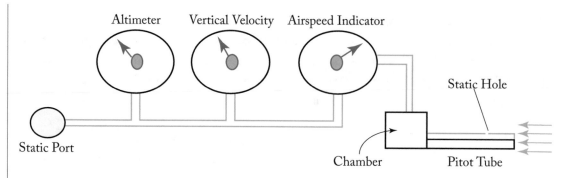

Figure 5.8: Pitot-static measurement device.

Due to safety reason and FAA regulations, all aircraft must use a device called Pitot tube to measure the aircraft speed. Airspeed is measured by comparing the difference between the pitot and static pressure (Figure 5.8) and, through mechanical linkages, displaying the resultant on the airspeed indicator. A static port (tube) only measures the static pressure, since the hole is perpendicular to the air flow, so the flow must turn 90° to enter into the tube. In contrast, a pitot tube measures the dynamic pressure, since the hole is facing the airflow. When a pitot tube, has a static port, it is often referred to as the pitot-static tube.

By employment of the pitot tube and the static port, we are dealing with three types of aircraft speeds (i.e., airspeed): (1) indicated airspeed, (2) true airspeed, and (3) equivalent airspeed. Before explaining the difference between these three terms, it is beneficial to introduce how a pitot tube works.

5.6.4 ALTITUDE SENSOR

Electronic sensors for measuring altitude and height, that is height above ground/sea-level, include those measuring distance by timing pulses of radio, laser or acoustic energy from transmission to return. These vary in their accuracy, depending upon their frequency and power. Radio altimeters vary in their accuracy and range depending upon their antennae configuration. Laser systems may have problems in causing eye damage and precautions must be taken in their selection and use.

Another altitude sensor is a mechanical device, is based on pressure, and is less accurate than electronic sensor. The pitot tube and static pressure holes are located at a suitable convenient position on the aircraft. Some convenient locations include: (1) under the wing, (2) at the middle of fuselage nose, and (3) beside fuselage front or middle section. The location of the static tapping is very important because it is essential to select a position where the local static pressure is the same as that in the free stream. The location of the pitot tube is also very important because it is essential to select a position where the local airspeed is the same as that in the free stream, and also is not too sensitive to change in the aircraft angle of attack and sideslip angle. The pitot and static holes are normally heated to avoid icing at low temperatures and high altitudes. The location of the holes will usually induce 2–5% error in reading, so the pressure difference measuring device must be calibrated.

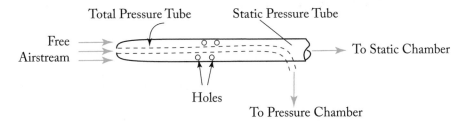

Figure 5.9: A pitot-static tube.

As an alternative to using separate pitot tube and static tapping, it is more convenient to use a combined device called the pitot-static tube (see Figure 5.9). The pitot-static tube consists of two co-centric tubes. The inner one is simply a pitot tube, but the outer one is sealed at the front and has a few small holes in the side. By mounting it on under the wing or at the fuselage side it can be arranged so that it is well clear of interference from the flow around the aircraft. Both pitot-static tube and separate pitot and static tapings do the same job for accurate speed measurement.

5.7   DESIGN CONSIDERATIONS

In the preceding sections, various navigation techniques have been described. In designing navigation system for a UAV, one must select the type of navigation system, select the navigation devices, sensors, and then conduct some calculations and analysis. In general, the primary criteria for the design of navigation system are as follows: (1) manufacturing technology, (2) required accuracy, (3) range, (4) weather, (5) reliability, (6) life-cycle cost, (7) UAV configuration, (8) stealth requirements, (9) maintainability, (10) endurance, (11) communication system, (12) aerodynamic considerations, (13) processor, (14) complexity of trajectory, (15) compatibility with guidance system, and (16) weight. Figure 5.10 shows the navigation design process as a subsystem of the autopilot.

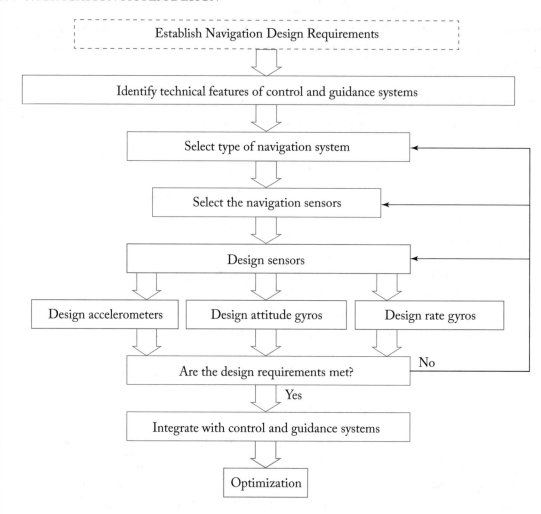

Figure 5.10: Inertial navigation system design process.

This figure illustrates a flowchart that represents the inertial navigation design process. In general, the design process begins with a trade-off study to establish a clear line between cost and performance (i.e., accuracy) requirements, and ends with optimization. The designer must decide about two items: (1) select type of navigation system and (2) select the navigation sensors. After conducting the calculation process, it must be checked to make sure that the design requirements are met. A very crucial part of the design process is to integrate the navigation system with control and guidance systems (i.e., a consistent autopilot). If a complete navigation system (e.g., GPS) is selected/purchased, the integration process must be still conducted. This includes activities such as matching frequencies, interfaces, and electric power requirements. More details are outside the scope of this work.

5.8 QUESTIONS

1. What is primary function of navigation system?

2. What are the two major sensors in inertial navigation?

3. What does IMU stand for?

4. What does GPS stand for?

5. Name three basic coordinate systems.

6. What is the function of an accelerometer?

7. What is the function of a rate gyroscope?

8. What is the function of a magnetometer?

9. Describe number of and the configuration of GPS satellites.

10. What is the process to convert the output signal of an accelerometer to a linear velocity?

11. What are the input and output parameters of navigation equations?

12. Describe the elements and the operation of an early mechanical accelerometer.

13. Describe the basic principle of working gyro.

14. What is the main difference between an attitude gyro and a rate gyro?

15. Describe the function of a roll-angle-hold autopilot.

16. What is the law of the gyro?

17. What are the name of three axes of a gyro?

18. Describe the celestial navigation.

19. Describe the inertial navigation system design process.

20. Name three activities included in the integration process of a navigation system.

21. What are the primary criteria for design of the navigation system?

CHAPTER 6

Guidance System Design

6.1   INTRODUCTION

Guidance is fundamentally defined as the process for guiding the path of a vehicle toward a given point. One of the primary functions of an autopilot is to guide the UAV to follow the trajectory. Most UAVs are planned to follow a predetermined and fixed trajectory. However, there are some classes of UAVs, mainly military ones, which their mission is to follow a moving target. In either case, the generation of a trajectory is the primary function of the guidance system. For the case of a fixed trajectory, the guidance system will be simple and straightforward. However, for the case of following a moving target, the autopilot requires a complex guidance system to generate the trajectory and to create the command for the control system. A UAV—through guidance and control systems—must generate the steering commands and subsequent control surface deflections to adequately adjust its way along the chosen flight path. Another important UAV operation where the guidance system is necessary is formation flight of multiple UAVs.

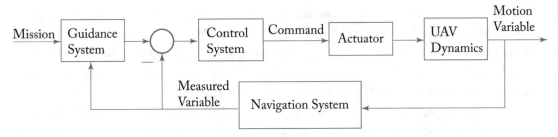

Figure 6.1: Control, guidance, and navigation systems in an autopilot.

Guidance is the process of producing a trajectory based on what is received from the command subsystem and the feedback from the navigation system. The guidance system is originally developed by the missile industries. Pastrick et al. [18] review the guidance laws for short-range tactical missiles. In a guided missile, the guidance system is only used in the homing phase of flight (last phase) to impact the target. For a full autonomous operation, the vehicle needs to have: (1) navigation, (2) guidance, and (3) control system of which the guidance system is the key element. The Predator A is originally designed to be flown in fully manual mode; so it did not require a guidance system. However, in later version, it had a guidance system such that it can be programmed to automatically return to base if the datalink is lost. Moreover, the Global Hawk (Figure 1.1) can be operated in a fully autonomous mode, thus it requires a guidance system.

Based on the guidance law, the guidance subsystem then produces the desired states which go to the control subsystem. The output of the guidance subsystem (Figure 6.1) is sent to the control system, and the control system implements this command through actuators driving control surfaces such as the elevator, aileron, and rudder. Guidance is mainly responsible for measuring and controlling the flight variables including the aircraft's angles, the rate of change of the angles, and the body axis accelerations. The navigation system then calculates the location of the aircraft, compares it with the pre-determined reference trajectory, and modifies the autopilot commands to drive the error to zero. The guidance subsystem often produces an acceleration command. Thus, the guidance subsystem makes the necessary correction to keep the vehicle on course by sending the proper command to the control system of autopilot.

For the case of UAVs, there are many cases where the guidance system is needed. A few examples are: (1) the current heading (Ψo) of the UAV does not lead to the desired position (Figure 6.2), (2) a UAV is seeking a fixed target; in order to make some measurements (e.g., take picture); (3) a UAV is following a moving ground target; and (4) a follower UAV is following a leader UAV; for a formation flight. One of the sub-items of case 1 is the "sense/detect and avoid" capability which means the capability of an unmanned aircraft to remain a safe distance from and to avoid collisions with other airborne aircraft. General Atomics Aeronautical Systems had the successful flight test of a prototype of its Due Regard Radar (DRR), an air-to-air radar system that supports GA-ASI's overall airborne Sense and Avoid architecture for Predator B (Figure 4.10). The main sensors of the Warrior (a variant of the MQ-1 Predator), are an SAR/GMTI (Synthetic Aperture Radar/Ground Moving Target Indicator) in the large fairing. The relation between the control system, the guidance system, and the navigation system is shown in Figure 6.1.

An important application of guidance system is the detect/sense-and-avoid a collision. This is and will be one of the biggest limitations to the widespread use of unmanned vehicles in civilian airspace. In manned civilian aviation, see-and-avoid is the primary mechanism by which piloted aircraft avoid collisions with each other. This is not practical for widespread use of unmanned vehicles, so they must achieve an equivalent level of safety to that of manned aircraft operations. There is currently a large amount of research being conducted on the UAV detect-and-avoid problem. Active solutions include the use of a sensor (e.g., radar) to detect collision threats, however this requires high amounts of electrical power, and is a bit heavy. Passive solutions include the use of machine vision (e.g., camera), which reduces the power requirement, however, has a high computational requirement. This chapter will describe various aspects of guidance system, elements of from guidance, guidance laws. In addition, three popular guidance laws are presented in details.

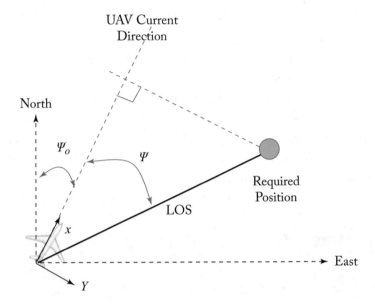

Figure 6.2: A case where a guidance system is necessary.

6.2   ELEMENTS OF GUIDANCE SYSTEM

The most important subsystem in a UAV compared with a manned aircraft is the autopilot. The autopilot is a vital and required subsystem within unmanned aerial vehicles. The autopilot has the responsibility to: (1) stabilize under-damped or unstable modes, (2) accurately track commands generated by the guidance system, (3) guide the UAV to follow the trajectory, and (4) determine UAV coordinates (i.e., navigation). Therefore, an autopilot consists of: (1) command subsystem, (2) control subsystem, (3) guidance subsystem, and (4) navigation subsystem. These four subsystems must be designed simultaneously to satisfy the UAV design requirements.

The guidance subsystem generates command for control system (i.e., generates suitable trajectories to be followed). It has mainly four elements: (1) the guidance measurement device (e.g., radar/seeker), (2) guidance law, (3) hardware (e.g., processor), and (4) software/programming. Chapter 7 covers microprocessors which is the main hardware for the guidance system.

The type of guidance system used depends upon the type and mission of the UAV being controlled, and they can vary in complexity from an inertial guidance system for long range UAV to a simple system where an operator visually observe the UAV (e.g., remote controlled aircraft) and sends guidance commands via a radio link. In any case, the guidance command serves as the input to the UAV control system. The command may be in the form of a heading or attitude command, a pitching or turning rate command, or a pitch or yaw acceleration command, depending upon the type of guidance scheme used.

Hardware equipment in the guidance system typically includes: homing head (or seeker), measurement device, transmitter, and processor. The homing head might be active, passive or semi-active. There are several guidance laws and several guidance system categories. The important point is to make sure that the guidance law matches the guidance system category. The guidance system categories includes: (1) Line Of Sight, (2) Navigation Guidance (e.g., Inertial Navigation, GPS), and (3) Homing (e.g., radar, infrared, and television). The typical guidance law includes: (1) collision, (2) proportional navigation guidance, (3) constant beam course, (4) pursuit guidance, (5) three point guidance, (6) optimal control guidance, (7) lead guidance, (8) lead angle, and (9) preset.

Table 6.1: Variables in four subsystems

No	Subsystem	Input	Output
1	Command subsystem	X, Y, Z	V_t, γ, ψ, R
2	Navigation subsystem	$U, V, W, P, Q, R, \theta, \phi, \psi$	X, Y, H, V_t
3	Guidance subsystem	Command: V_t, γ, ψ Output: $H, \Delta Y, \Delta \psi$	$\alpha, \gamma, \phi, \beta$
4	Control subsystem	Desired: $\alpha, \gamma, \phi, \beta$ Output: $V, P, Q, R, \beta, \phi, \alpha, \gamma$	$\delta_T, \delta_E, \delta_A, \delta_R$
5	UAV dynamic model	$\delta_T, \delta_E, \delta_A, \delta_R$	$\alpha, \gamma, \phi, \beta, V, Q, P, R$

An important UAV operation where the guidance system is necessary is formation flight. A simple design method is preferable for such mission. One benefit of this approach is that the guidance and flight control design process is integrated. There is a possibility that the guidance law becomes complex as the number of UAVs is high. Table 6.1 represents the state and control variables that each subsystem receives and produces.

In general, the primary criteria for the design of guidance system are as follows: (1) manufacturing technology, (2) required accuracy, (3) range, (4) structural stiffness, (5) load factor, (6) weather, (7) maneuverability, (8) reliability, (9) life-cycle cost, (10) UAV configuration, (11) stealth requirements, (12) maintainability, (13) endurance, (14) communication system, (15) aerodynamic considerations, (16) processor, (17) complexity of trajectory, (18) compatibility with control system, (19) compatibility with navigation system, and (20) weight. More details on this subject are out of scope of this text.

6.3   GUIDANCE LAWS

In a conventional autopilot, three laws are governing simultaneously in three subsystems: (1) control system through a control law, (2) guidance system via a guidance law, and (3) navigation system through a navigation law.

In the design of an autopilot, all three laws need to be determined/designed. The determination of these laws is at the heart of autopilot design process. A guided UAV is guided according to a certain guidance law. The guidance subsystem generates command signal for the control system. The guidance command is usually generated in a very short period of time. In practice, it can be assumed to be continuously generated. The guidance law is the basis for generation of guidance signal.

Guidance is the means by which a UAV steers, or is steered, to a target. A fully autonomous UAV is guided according to a certain guidance law. We may classify guidance laws as classical and modern guidance laws. The modern guidance laws are derived from optimal control theory, differential games, and singular perturbation theory.

There are various types of guidance laws. In this section, three more effective and popular ones are presented. Three classical guidance laws are: (1) LOS guidance (i.e., command guidance), (2) proportional navigation guidance, and (3) pursuit guidance. The classical guidance laws are employed in missiles for decades and are designed using rather simple ideas. Of these, the proportional navigation (PN) guidance laws form the boundary between the classical and the modern approach. These laws each have various variants, which are not reviewed here.

The selection of guidance law dictates the selection of hardware, and influences the overall cost. Moreover, the type of guidance law will affect the control system, and navigation system. Thus, a systems engineering vision is required to select the best guidance law for a specific mission. It is the guidance law that, in principle, distinguishes an unguided projectile from a guided UAV. A majority of available guided missiles use these guidance laws or their variants. They have the advantage of standard equipment, easy mechanization, and minimal information requirement. Their disadvantage lies in the fact that their accuracy suffers against maneuvering and intelligent targets. The design and analysis of guidance laws has been an active area of research for the past seven decades.

Here, a short description of these three law is provided. In future sections, the details are presented.

1. The basic principle in line-of-sight (LOS) guidance law is to guide the UAV on a LOS course in an attempt to keep it on a line joining the target and the ground station (tracking line).

2. However, for proportional navigation guidance law, the UAV is commanded to turn at a rate proportional to the angular velocity of the line of sight (LOS).

3. The basic idea for the pursuit guidance law is to keep the UAV pointed toward the target.

Blakelock [16] compares classical guidance laws, and provides details for modern guidance laws. Among these three laws, the proportional navigation is the most popular. The advantage of proportional navigation guidance law lies in the fact that it is easy to mechanize, and requires easily obtainable information.

6.4   LINE-OF-SIGHT GUIDANCE LAW

The basic principle in line-of-sight (LOS) guidance law is to guide the UAV on a LOS course in an attempt to keep it on a line joining the target and the ground station (tracking line). Of three cases where a UAV requires a guidance system, the case of a formation flight is examined; where, there is a leader UAV and a number of (say four) follower UAVs (see Figure 6.3). For such a case, the line of sight is defined as the line joining the follower UAV and the leader UAV. In addition, the leader UAV is following a moving ground target. For this law, the target-tracking radar acquires the target shortly after take-off and then guides the UAV into the beam of the target-tracking radar. For the guidance command the actual distance from the tracking line to the UAV is required.

In the LOS law, the velocity of the follower UAV perpendicular to the LOS should be equal to the LOS velocity at that point. It is assumed that the LOS variables are available from the use of onboard vision sensors.

$$V_{Fn} = D_{FL} \dot{\Psi}_{LOS} \tag{6.1}$$

where $\dot{\Psi}_{LOS}$ is the rate of change of heading angle of the line of sight, and D_{FL} denotes the distance between the follower UAV and the leader UAV. Moreover, V_{Fn} is velocity of the follower UAV perpendicular to the *LOS*. An imaginary line between the follower UAV to the leader UAV is referred to as line-of-sight. The LOS guidance scheme can be mechanized in two ways: command line-of-sight (CLOS), and beam rider (BR) guidance scheme. In CLOS guidance scheme there is an uplink to transmit guidance signal from a ground station to the flower UAV. Hence, it is required by the ground station to track the UAV as well as the Leader UAV and target. So, the follower UAV is guided so as to remain on the commanded LOS.

In the BR guidance scheme, an electro-optical beam is directed at the leader UAV from follower UAV. There are sensors inside the follower UAV which senses the deviation of the follower UAV from the centerline of the beam. The autopilot of the follower UAV generates appropriate guidance commands to annul this deviation. Thus, the follower UAV is required to constantly track the leader UAV. Performance of UAVs using LOS guidance is good in a low maneuver formation flight at short ranges and moderate speed. However, these UAVs suffer from certain disadvantages. A major disadvantage is that the commanded speed required for approaching leader UAV becomes very high toward the end, since UAV has limit for max speed. In addition, the performance degrades against high-speed and maneuvering targets.

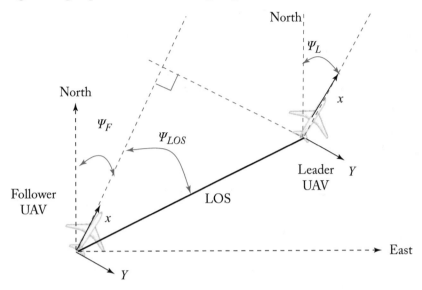

Figure 6.3: Line-of-sight (top view).

6.5   FORMATION FLIGHT

Formation flight is the orchestrated flight of more than one aircraft. This type of flight originates from manned aircraft, and FAA regulates its minimum legal requirements. There are various arrangement; a few examples are: (1) parade formation, (2) standard formation turn, and (3) echelon turn. In order to prevent any collision in formation flight, the pilot should always keep all airplanes in front in sight, and be able to pass behind and under any airplane in front on your way to the outside.

One beautiful example of formation flight in nature is the murmuration which is a phenomenon of nature that amazes and delights those lucky enough to witness it. Is this event, hundreds, even thousands, of starlings flying together in a whirling, ever-changing pattern. As they fly, the starlings in a murmuration seem to be connected together. They climb, descent, and turn and change direction at a moment's notice.

With advances in science and technology, UAVs are able to conduct formation flight. One of difficult parts of formation flight is for a UAV to joining a group of UAVs. The appropriate guidance law plus a precision control may guarantee the safety of formation flight. To maintain the format and a safe distance, there are a number of leader UAVs, and a number of follower UAVs. There are no unique standard formation flight, however, in each case, the follower UAV must maintain a lateral, longitudinal, and a vertical distance from the leader UAV.

Now, as an example, consider four unmanned aerial vehicles are required to circle over a moving target (Figure 6.4) with extreme accuracy, in order to gather intelligence and transmit information instantly back to ground station. In the formation flight system considered here, each ith aircraft uses information from itself and the preceding (*i-1*)*th* aircraft to track a commanded relative position. When all four UAVs entered the target circle, each UAV is a leader as well as a follower UAV. Because each UAV is required to maintain a coordinated turn around the target, it will be assumed to be a leader UAV. In addition, since each UAV needs to maintain a fixed spatial distance from another leading UAV, each UAV is assumed to be a follower as well. Thus the integrated guidance and control system of all four UAVs are similar. The relation between the control system, the guidance system, and the navigation system for each UAV is shown in Figure 6.5.

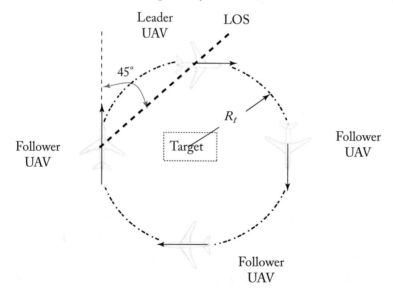

Figure 6.4. Leader-follower UAV geometry.

Since the flight path is a circle, and there are four UAVs, the LOS or tracking line must always make 45° with the follower heading angle. When the follower UAV is turning around the target, the spatial distance with the leader UAV is achieved using the engine throttle. When the LOS angle (Ψ_{LOS}); the angle between the follower UAV heading angle ((Ψ_F) and the leader UAV heading angle (Ψ_L) is more than 45°, the throttle is deflected such that to increase the engine thrust and accelerate the UAV. Based on a desired circular path, the following relationships may be derived by observation:

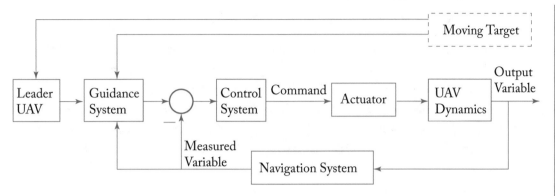

Figure 6.5: Control, guidance, and navigation systems in the follower UAV autopilot.

$$\Psi_F - \Psi_L = 90° \tag{6.2}$$

And, the light of sight angle is:

$$\Psi_{LOS} = 45° \tag{6.3}$$

In case that the angle between LOS and the follower heading angle is less than 45°, the throttle is deflected such that to decrease the engine thrust and decelerate the UAV. As soon as the follower UAV is reached to the commanded circle around the target and stabilized, the guidance system will be activated to guide the aircraft such that to keep a constant 45° of LOS angle. When the follower UAV's LOS is different than the commanded LOS, the guidance system generates a yaw rate (R) and a change in the aircraft speed for control system:

$$R = k(\Psi_{LOS} - \Psi_{LOS_C}) \tag{6.4}$$

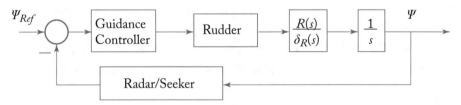

Figure 6.6: Block diagram of the LOS guidance system.

The constant k is determined in design process of the guidance law. For this objective, the knowledge of transfer function between rudder deflection (δ_R) to the yaw rate (R); or ($R(s)/\delta_R(s)$) is required. A general block diagram of the LOS guidance plus control systems is depicted in Figure 6.6.

6.6   PROPORTIONAL NAVIGATION GUIDANCE LAW

Proportional navigation (PN) is known to be the most widely used guidance law in UAVs and modern missiles. Please note that proportional navigation has nothing to do with navigation, it is just a guidance law used to guide air vehicles. This misnomer dates back to the early days of navigation by ships. In proportional navigation guidance law, the UAV is commanded to turn at a rate proportional to the angular velocity of the line of sight (LOS). The ratio of the turn rate (ω) to the angular velocity of the LOS ($\dot{\Psi}_{LOS}$) is called the proportional navigation constant, N.

$$N = \frac{\omega}{\dot{\Psi}_{LOS}} \tag{6.5}$$

The constant N is always greater than unity, and usually ranges from 1.5–4. This means that the follower UAV will be turning faster than the LOS, and thus the UAV will build up a lead angle with respect to the LOS. The goal is to reach a keep desired difference between the heading angle of follower UAV and the leader UAV or target. If $N = 1$, then the follower UA is turning at the same rate as the LOS, or simply following on the target. The direction of the LOS is stablished by the follower UAV seeker, in tracking the target. The component of UAV velocity perpendicular to the LOS (V_{Fn}) is:

$$V_{Fn} = V_F \sin[\Psi_{ref} - (\Psi_F - \Psi_L)] \tag{6.6}$$

which generates a positive LOS rotation. If the perpendicular components of follower UA and target velocity are equal and unchanging, there will be no rotation of the LOS. When the distance between the UAV and target is kept constant, the objective of guidance is met. Thus in the steady state, the turn rate of the follower UAV is:

$$\omega_F = N\Psi_{LOS} \tag{6.7}$$

This command will be fed to the control system. The block diagram for his guidance system (plus control system) is identical to Figure 6.6, the difference is the guidance controller, which is working on PN law. Recall that the objective is not to hit the target; it is to follow the target/leader UAV. The LOS rate is obtained by measuring the rate of rotation of the UAV seeker tracking the target.

6.7   PURSUIT GUIDANCE LAW

The basic idea in pursuit guidance is to keep the UAV pointed toward the target (either a ground target or a leader UAV). Whenever there is an undesired angular deviation, a command is generated to annul the deviation. There is a further classification of this guidance law as a. pure pursuit, and b. deviated pursuit. Pure pursuit makes the UAV point at the target while deviated pursuit makes the UAV point at a spot ahead of the target by a fixed angle. The idea behind deviated pursuit is to take advantage of the information of the target's direction.

In the pursuit guidance, the UAV longitudinal axis; attitude (or the UAV's velocity vector) is kept pointed at the target (note that, there is usually a non-zero angle of attack in UAVs). Figure 6.7 illustrates the trajectory of a UAV with a pursuit guidance against a moving target.

6.8   WAYPOINT GUIDANCE

In general, a waypoint is a term used to refer to an intermediate point or place on a route or line of travel, a stopping point or point at which course is changed. Waypoints are sets of coordinates that identify a point in physical space. GPS systems are increasingly used to create and use waypoints in navigating UAVs. GPS receivers used in aviation have databases which contain named waypoints, radio navigation aids, and airports.

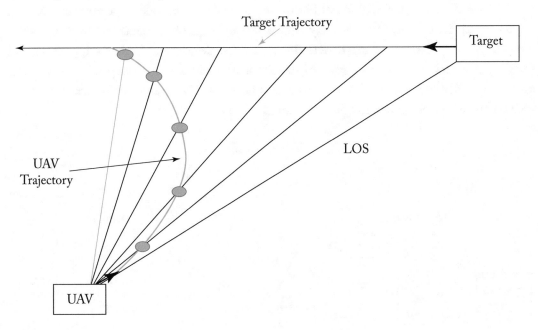

Figure 6.7: Pursuit guidance against a moving target.

A simple guidance technique for a UAV to follow a pre-planned trajectory is to follow waypoints. It is a route and destination planner for the UAV. This approach instructs the UAV where to fly, at what height, the speed to fly at, and it can also be configured to hover at each waypoint. Waypoint GPS Navigation guidance allows a UAV to fly on its own with its flying destination or points pre-planned and configured into the remote control navigational software. When waypoints are programmed into the application, the UAV can go directly to the first point and proceed to each point in turn. Thousands of waypoints can easily be programmed in a UAV software. In aviation, waypoints consist of a series of abstract GPS points (x, y, and z) that create artificial airways (i.e., highways in the sky). The coordinates of the way-points can be provided either before or after take-off. In this approach, the instruction is continually commanded to the control system to fly on a selected heading at a selected airspeed and altitude.

6.9   SEEKER

UAVs have proven to be useful for observing activity on the ground without placing an air crew at risk. An important element in such activity a UAV is the sensor. Some sensors are part of the payloads; while others are necessary for control/navigation/guidance systems. For instance, the al-timeter radar is a sensor and is employed for navigation purpose; while a camera is a sensor which configured as a payload. Ina addition, some radars are utilized as communication device.

Some radars are employed for visibility of ground activity, while some are utilized for seek-ing the target. Large UAVs usually carry optical sensors (e.g., TV cameras and forward looking infrared) which are less susceptible to detection than active devices such as radars. Under favorable conditions, optical sensors can supply high-quality images of the ground targets. However, opti-cal sensors suffer from a limited field of view and from severely reduced performance in adverse weather and battlefield smoke and dust conditions.

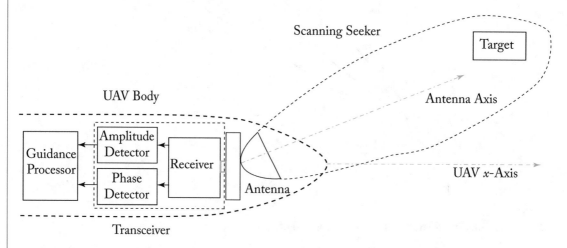

Figure 6.8: Configuration of radar, antenna, and guidance system

A seeker is a significant part of the guidance system where the UAV is required to detect and follow a target. The seeker, by tracking the target, establishes the direction of the LOS, and the output of the seeker is the angular velocity of the LOS with respect to inertial space as measured by rate gyros mounted on the seeker. The LOS rate is obtained by measuring the rate of rotation of the UAV seeker tracking the target. Basically, the main function of the UAV seeker (in the missile terminology, it is known as homing eye) subsystem is to:

1. provide the continuous measurements of target position and velocity,

2. track the target with the antenna, and

3. measure the line-of-sight angular rate.

The antenna of the seeker is of various types; the most popular ones are: (1) radar, (2) infrared (IR), (3) laser, and (4) electro-Optic (EO) (e.g., television camera). The UAV radar system can detect and track any moving targets/vehicles such as cars, tanks, trucks, and low-flying helicopters. There are three kinds of seeker radar: (1) active (which has both transmitter and receiver), (2) passive (which only has a receiver), and (3) semi-active (which is using the signal from ground station). The way in which closing velocity and LOS rate data is obtained to mechanize guidance laws is a function of the type of seeker that is used; and how it is mounted to the UAV body.

An active seeker radar detects a target by sensing electromagnetic energy reflected from the target's surface. The source of the reflected radiation is a radar transmitter. The transmitter must beam electromagnetic radiation at the target, this radiation must travel to the target, reflect, travel back to the receiving antenna of the UAV. It is then amplified, demodulated and analyzed to determine the direction of the target. This information enables the guidance computer to steer the UAV toward the target. An effective seeker must have the ability to discriminate between the target's return and reflections from its background, i.e., the surface of the ground or see. It should also be capable of resisting jamming or deception and be able to penetrate through adverse weather conditions. Figure 6.8 depicts the configuration of radar, antenna, and guidance system in a UAV. Figure 6.9 illustrates an MQ-4C UAV configuration with two seekers. Radar theory is a subject requiring a knowledge of electromagnetism and wave theory. This topic is beyond the scope of this text, the interested reader is referred to references such as Ben-Asher and Yaesh [21], Zarchan [22], and Palumbo et al. [23].

A typical UAV radar system can detect and track moving tactical vehicles such as tanks, trucks, and low-flying helicopters out to a range of 15 km. It has a 360° wide-area surveillance mode in which the radar antenna sweeps out an annular 5–15 km range swath every 18 sec [53]. At an operational altitude of 3 km, the swath corresponds to depression angles of 11°–37°, thus affording excellent visibility of ground activity.

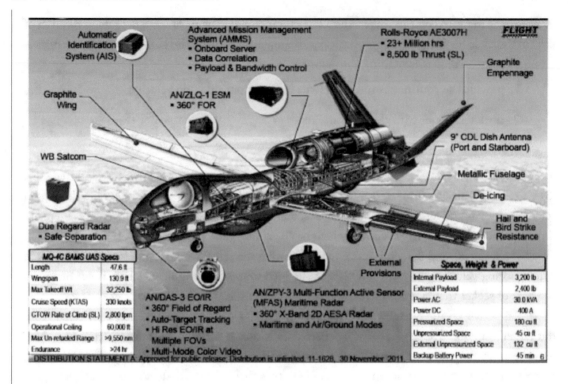

Figure 6.9: MQ-4C UAV configuration (with two seekers)

6.10   QUESTIONS

1. Define guidance system.

2. Define guidance law.

3. Describe PN guidance law.

4. Describe pursuit guidance law.

5. Describe line-of-sight guidance law.

6. Name the main elements of a guidance system.

7. What is the main function of a seeker?

8. What are the three classical guidance laws?

9. What is the most widely used guidance law?

10. What requirements will influence the design of guidance system?

11. Define line of sight.

12. What are the functions of a seeker?

13. List four types of seeker.

CHAPTER 7

Microcontroller

7.1 INTRODUCTION

The autopilot is as an essential component of a UAV to execute the automatic flight control activities. Four major autopilot functions are: (1) guidance, (2) navigation, (3) control, and (4) tracking. An autopilot should be able to: (1) measure flight parameters, (2) process data, (3) make decisions, and (4) create commands. All of these activities are performed via a microcontroller. The heart of a UAV autopilot is the microcontroller. A microcontroller is essentially an integrated circuit (IC) that is programmed to a do a specific task. By reducing the size and cost compared to a design that uses a separate microprocessor, memory, and input/output devices, microcontrollers make it economical to digitally control even more devices and processes. The majority of microcontrollers in use today are embedded in devices and machinery equipment. Microcontrollers are used in automatically controlled products and devices, such as automobile cruise control; implantable medical devices; remote controls; office machines; appliances; power tools; toys; automatic door opener; and remotely controlled airplanes.

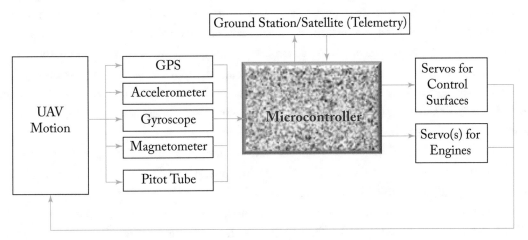

Figure 7.1: Microcontroller connections (inputs/outputs).

Microcontrollers must provide real-time response to events in the embedded system they are controlling. A customizable microcontroller incorporates a block of digital logic that can be personalized for additional processing capability. Microcontrollers must usually have a low-power requirement (0.5–1 Watt); since many devices they control are battery-operated. Microcontrollers are designed for embedded applications, in contrast to the microprocessors used in personal computers or other general purpose applications consisting of various discrete chips. A microcontroller [31] can be considered a self-contained system with a processor, memory, and peripherals and can be used as an embedded system. Integrating all elements on one chip saves space and leads to both lower manufacturing costs and shorter development times.

Figure 7.1 illustrates a typical microcontroller connections (inputs/outputs) in an unmanned aerial vehicle. It receives signals from devices such as gyroscope, and sends signals to elements such as engine. Some microcontroller designers/manufacturers/architectures/vendors are: Intel, Microchip, ARM, Texas Instrument, Ardupilot (e.g., see Figure 7.2), Toshiba, Philips, Atmel, Siemens, PIC, and Motorola.

This chapter presents the fundamentals of microcontroller; and how to setup a microcontroller to perform various functions of an autopilot. The level of complexity of a microcontroller depends on the UAV mission, and the equipment to be controlled during a flight operation. For instance, when the mission range is longer than the line of sight, there is a need for sophisticated radars and powerful CPUs. As the UAV range and endurance are increased, the microcontroller requires more supporting equipment, with a higher level of complexities.

7.2   BASIC FUNDAMENTALS

A microcontroller is an electronic device with several elements to allow a UAV to monitor the flight, and control its flight operation. The common features of a microcontroller is briefly described in this section. The peripherals of a microcontroller are integrated into a single chip, so, the overall weight and cost is very low. It has a central processing unit (CPU), in-circuit programing, and in-circuit debugging support. It is also equipped with Analog-to-Digital (A/D) converters, some include Digital-to-Analog (D/A) converters. For program, data storage, and operating parameters storage, various types of memories such as RAM[7], ROM[8], and flash memory are utilized.

For the purpose of system interconnect, various interfaces and serial input/output such as serial ports, Inter-Integrated Circuit (I^{2}C), serial peripheral interface, and controller area network are installed. Most microcontrollers are also have peripherals such as timers, event counters, PWM generators, and watchdog clock generator. Other possible items are: (1) power supply, (2) USB interface, (3) pin header, (4) LCD, (5) buttons, and (6) potentiometers. The size and cost of a microcontroller is a function of computing needs, e.g., speed, capacity of memory, number of I/O[9] ports, power consumption, and timers. The microcontrollers are easily upgradable. Figure 7.2

---

[7] Random Access Memory
[8] Read Only Memory
[9] Input/Output

demonstrates an Arduino microcontroller board with its major elements. The input and output pins are at the top and bottom. A 28-pin microcontroller is located in the lower center.

A microcontroller [30] is a small computer on a single integrated circuit containing: (1) processor core, (2) memory, (3) programmable input/output peripherals, and (4) code. A microcontroller is a small, low-cost computer-on-a-chip; it is often used to run dedicated code that controls one or more tasks in the operation of a device or a system. They are called embedded controllers because the microcontroller and support circuits are often built into, or embedded in, the devices they control.

A microcontroller has many similarities and differences with a microprocessor. A computer requires a processor/microprocessor to support myriad computing functions. In both microprocessor and a microcontroller, algorithms must be executed on digital computation devices. Table 7.1 [31] illustrates a comparison between a microprocessor and a microcontroller. Microprocessors available to aerospace generally lag the performance available for ground-based computing. Microprocessors are an integral part of autopilots, INS[10], VMS (Vehicle Management System), device controllers, and even sensors.

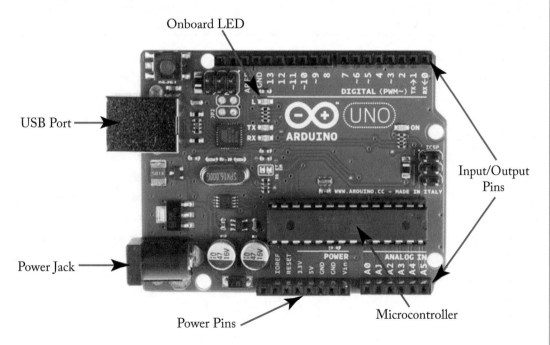

Figure 7.2: An Arduino microcontroller board (image courtesy of Arduino.cc).

---

[10] Inertial Navigation System

Microcontrollers have various packaging formats: (1) dual inline package (DIP) through hole (8 pins); (2) small Output IC (SOI) with surface mount (18 pins); (3) quad flat package (QFP) with surface mount (32 pins); and (4) ball Grid Array (BGA) with surface mount (100 pins).

Microcontrollers have embedded systems which usually have no keyboard, screen, disks, printers, or other recognizable I/O devices of a personal computer, and may lack human interaction devices of any kind. Microcontrollers may not implement an external address or data bus as they integrate RAM and non-volatile memory on the same chip as the CPU. Using fewer pins, the chip can be placed in a much smaller, cheaper package. Many embedded systems need to read sensors that produce analog signals. This is the purpose of the analog-to-digital (A/D) converter.

Table 7.1: A comparison between a microprocessor and a microcontroller

No	Criterion	Microprocessor	Microcontroller
1	Cost	High	Low
2	Energy use	Medium to high	Very low to low
3	Applications	General computing (e.g., laptops, computers)	Appliances, specialized devices
4	Speed	Very fast	Relatively slow
5	External parts	Many	Few

Microcontrollers usually contain from several to dozens of General Purpose Input/Output pins (GPIO). The GPIO pins are software configurable to either an input or an output state. When GPIO pins are configured to an input state, they are often used to read sensors or external signals. Configured to the output state, GPIO pins can drive external devices such as LEDs or motors, often indirectly, through external power electronics.

In many cases, a microcontroller communicates bi-directionally with the wireless telemetry system which relays data back to a ground station. This allows user to receive real time data about the aircraft, and be able to update commands or change the mission parameters on-the-fly. A remote control input is a signal that can be passed through the microcontroller to the servos and speed controllers if the user wants to manually override the flight controller with a remote control transmitter and the aircraft is within range. Microcontrollers may be classified from various aspects including bits, memory, instruction set, and architecture. Figure 7.3 illustrates various types [33] of microcontrollers.

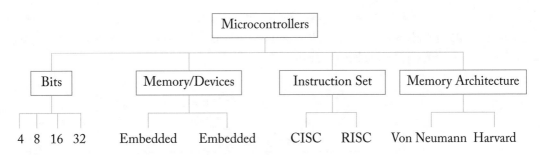

CISC: Complex Instruction Set Computer
RISC: Reduced Instruction Set Computer[11]

Figure 7.3: Types of microcontroller.

Harvard Architecture: This architecture demands that program and data are in separate memories which are accessed via separate buses. In consequence, code accesses do not conflict with data accesses which improves system performance. As a slight drawback, this architecture requires more hardware, since it needs two busses and either two memory chips or a dual-ported memory (a memory chip which allows two independent accesses at the same time).

Von Neumann Architecture: In this architecture, program and data are stored together and are accessed through the same bus. This implies that program and data accesses may conflict (resulting in the famous von Neumann bottleneck), leading to unwelcome delays.

A UAV designer has at least four options for number of bits (4, 8, 16, or 32)—two options for type of memory (embedded versus external); two options for instruction set (CICS versus RICS)—and two types of memory architecture (Von Neumann vs. Harvard). The main criterion for selecting the best option is to weigh cost versus performance. The optimum selection may be determined through an optimization process. Since this chapter covers the fundamentals of microcontroller, having at least one computer science engineer is a must in the team of UAV design/development.

7.3 MODULES/COMPONENTS

Figure 7.4 demonstrates a basic layout [35] of a microcontroller and typical microcontroller connections (inputs/outputs). The main modules are processor core, SRAM, EEPROM[12], time, digital I/O module, serial interface, analog module, and interrupt module. These modules typically found in a microcontroller. Typical microcontroller input and output devices include: (1) switches, (2) relays, (3) solenoids, (4) LEDs, (5) small or custom liquid-crystal displays, (6) radio frequency devices, and (7) sensors for data such as temperature, humidity, light level. A number of measurement

---

[11] Microcontrollers are mostly RISC.
[12] Electrically Erasable Programmable Read Only Memory.

devices such as pitot tube, GPS, gyroscope, magnetometer, altimeter, accelerometer, and rate gyro are communicating with the microcontroller via interfaces and modules. In this section, the major modules of a microcontroller are briefly described.

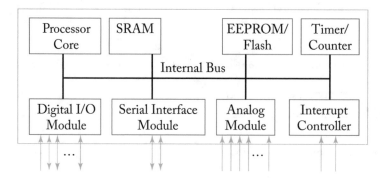

Figure 7.4: Basic layout of a microcontroller.

1. **Processor Core:** The processor core (CPU) is the main part of any microcontroller. It contains the arithmetic logic unit, the control unit, and the registers (stack pointer, program counter, accumulator register, register file, etc.). Computers only use 0 and 1 to represent numbers and letters.

2. **Memory:** A memory is an element to store information. The memory is split into program memory and data memory. In larger controllers, a DMA controller handles data transfers between peripheral components and the memory.

3. **Timer/Counter:** The timer/counter is used to timestamp events, measure intervals, and count events. Most controllers have at least one timer/counter. Timers are used for a variety of tasks ranging from simple delays over measurement of periods to waveform generation. The most basic use of the timer is in its function as a counter. Timers generally allow the user to timestamp external events, to trigger interrupts after a certain number of clock cycles, and even to generate pulse-width modulated signals for motor control. Many controllers also contain PWM[13] outputs, which can be used to drive motors. Furthermore, the PWM output can, in conjunction with an external filter, be used to realize a cheap digital/analog converter. Each timer is basically a counter which is either incremented or decremented upon every clock tick.

4. **Digital I/O:** A signal is either digital (combinations of 0 and 1) or analog (such as voltage). The microcontroller digitizes an analog value by mapping it to one of two states, logical 0 or logical 1. The analog and digital signals are transferred differently.

[13] Pulse Width Modulation.

Parallel digital I/O ports are one of the main features of microcontrollers. The ability to directly monitor and control hardware is the main characteristic of microcontrollers. As a consequence, practically all microcontrollers have at least 1–2 digital I/O pins that can be directly connected to hardware. The number of I/O pins varies from 3 to over 90, depending on the controller family and the controller type. I/O pins are generally grouped into ports of 8 pins, which can be accessed with a single byte access.

5. **Analog I/O:** The analog I/Os are the way to transfer analog signals. Apart from a few small controllers, most microcontrollers have integrated analog/digital converters, which differ in the number of channels and their resolution (8–12 bits). The analog module also generally features an analog comparator. In some cases, the microcontroller includes digital/analog converters.

6. **Interrupt Controller:** Interrupts are useful for interrupting/stopping the normal program flow in case of (important) external or internal events. In conjunction with sleep modes, they help to conserve power.

7. **Interfaces:** Controllers generally have at least one serial interface which can be used to download the program and for communication with the user. Serial interfaces also are used to communicate with external peripheral devices; most controllers offer several and varied interfaces. Many microcontrollers also contain integrated bus controllers for the most common busses. Larger microcontrollers may also contain USB or Ethernet interfaces.

The basic purpose of any interface is to allow the microcontroller to communicate with other units, (e.g., other microcontrollers, peripherals, or a host computer). The implementation of such interfaces can take many forms, but basically, interfaces can be categorized according to a handful of properties: They can be either *serial* or *parallel*, synchronous or asynchronous, use a bus or point-to-point communication, be full-duplex or half duplex, and can either be based on a master-slave principle or consist of equal partners.

8. **Watchdog Timer:** Since safety-critical systems form a major application area of microcontrollers, it is important to guard against errors in the program and/or the hardware. The watchdog timer, also sometimes called COP[14] is used to monitor software execution. The basic idea behind this timer is that once it has been enabled, it starts counting down. When the count value reaches zero, a reset is triggered, thus reinitializing the controller and restarting the program. To avoid this controller reset,

[14] Computer Operates Properly.

the software must reset the watchdog before its count reaches zero. The watchdog timer is used to reset the controller in case of software "crashes".

9. **A/D and D/A Converter:** Since processors are built to interpret and process digital data (i.e., 1s and 0s), they are not able to do anything with the analog signals that may be sent to it by a device. So the analog to digital converter is used to convert the incoming data into a form that the processor can recognize. A less common feature on some microcontrollers is a digital-to-analog (D/A) converter that allows the processor to output analog signals or voltage levels. The digital-to-analog conversion is a prerequisite for some analog-to-digital converters. Microcontrollers often have little or no analog output capabilities. So if the application requires a D/A converter, most of the time it has to be fitted externally.

10. **Debugger Unit:** Some controllers are equipped with additional hardware to allow remote debugging of the chip from a land computer. So there is no need to install a special debugging software, which has the distinct advantage that erroneous application code cannot overwrite the debugger.

A UAV designer has various options for selecting modules. The main criterion for selecting the best option is to weigh cost versus performance. The optimum selection will be determined through an optimization process.

7.4 FLIGHT SOFTWARE

7.4.1 SOFTWARE DEVELOPMENT

UAVs are software-intensive systems. A solid understanding of flight software is essential for success of a UAV program. Many UAV development programs reach the state where a nearly completed unmanned aircraft awaits software before flight test can begin. Much of the software that was developed by the UAV designer in the past is now developed by suppliers. UAV software is becoming a commodity for flight controls, data management, and ground control station software. The software development scope is underestimated, and often is not well understood. This section presents the requirements of an acceptable flight software as well as the microcontroller code. A UAV program requires a number of software to develop, including: (1) flight simulation software, (2) ground station software, (3) mission management system software, and (4) microcontroller software. This section introduces only microcontroller software.

7.4.2   OPERATING SYSTEM

The Operating System (OS) application programming interface (API) is a significant consideration, not only for execution, but also for ease of system development. Due to the critical nature of flight control, high reliability and real-time execution is required. A portable operating system interface is a recommended. This it is widely supported, and allows easy porting of applications programmed either to attempt for some fixed period of time to re-establish communications, to execute a fully automated operation, or to independently complete the mission.

7.4.3   MANAGEMENT SOFTWARE

An unmanned aircraft management system (UMS) performs flight critical and unmanned aircraft operation functions. This includes interfacing to command and control communication systems, subsystems, and the autopilot. Frequently, high-level flight route management occurs on the UMS ahead of the autopilot. The UMS is hosted on one or more computers. However, a mission management system (MMS) manages mission critical functions. These can include commanding payloads, selecting data feeds for downlink, data storage, data retrieval, data fusion, and payload downlink management. An architecture that segregates the MMS from the UMS and other flight critical systems is generally desirable for improved reliability. The MMS is hosted on one or more computers.

Large defense contractors can choose to specialize in mission system software, which runs on the MMS and the ground control station. The VMS efforts involve subsystem management and integration of various avionics components provided by multiple suppliers. The contingency management algorithms usually reside on the VMS as well.

The software that estimates the unmanned aircraft state based on sensor data generally runs on a procured device. The INS contains the inertial sensors and GPS, and all of the Kalman filtering is performed on the INS. The INS provides state data to the data bus, which is routed to the data consumers. Many autopilots have built in INS as well as air data sensors and include the necessary filters to estimate the state.

Ground payload software functionality: (1) ability to command the payload,(2) process payload state feedback, and (3) process the payload's collected data to generate usable intelligence. Ground payload software functions can be performed at: (1) MPO workstation, (2) separate mission management processors, and (3) other systems.

7.4.4   MICROCONTROLLER PROGRAMING

Microcontroller programs must fit [34] in the available on-chip memory, since it would be costly to provide a system with external, expandable memory. In general, microcontroller programming procedure is as follows.

- Write a code (in a language) for the microcontroller in an integrated development environment, a PC program.

- Debug the code.

- Compile the code into binary code which the microcontroller can execute.

- A programmer (a piece of hardware, not a person) is used to transfer (load) the code from the computer to the microcontroller.

Using third-party software can save nonrecurring engineering cost, but the UAV designer gives up some control. This supplier-provided software is sometimes called firmware. The UAV prime contractor can focus its attention on the unmanned aircraft management system mission management system.

Flight software should be reliable and predictable. Software that responds predictably to a set of inputs is known as deterministic. Aerospace software standards and certification methods help ensure solid design practices and software testing methods. Many government contracts require that the software developers or prime contractor achieve the certification.

Flight control software can run on an autopilot that includes the sensors or on a separate processor. The UAV prime contractor can implement the control laws themselves, though this work is increasingly performed by autopilot suppliers. Every device must be commanded, receive data, or provide data. This data exchange occurs over a real-time operating system (RTOS).

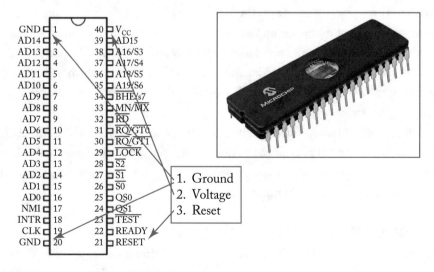

Figure 7.5: ISP output connections to an 40-lead PDIP (image courtesy of shannonstrutz.com).

The most common type of the programmer is an ICSP (In-Circuit Serial Programmer). An example is 8051 loader. ISP programmer have two functions—converts the code to binary and loads the code from the PC to the microcontroller. Common data bus types are: Ethernet; MIL-STD 1760; MIL-STD 1553; and CAN. Figure 7.5 shows an ISP output connections to an 40-lead PDIP.

7.4.5   SOFTWARE INTEGRATION.

Most payloads require interfaces with both the UA and ground-station software. Payloads can be an integral part of the system, such as is common with EO/IR surveillance payloads, where the payloads are nearly always on the UAV. The payload workstation is baseline in the ground control station. Most operational UASs are required to adapt to specialized payloads that might only be required for a small portion of the missions. Specialized payloads require adaptation of the system software—both on the air and ground.

Sophisticated UAVs have a distinct mission management system that interfaces to the payloads, data storage devices, and payload return link. Robust mission management architectures are segregated from flight-critical UAV systems. Dedicated on-chip hardware often includes capabilities to communicate with other devices (chips) in digital formats such as Inter-Integrated Circuit (I^{2}C), Serial Peripheral Interface (SPI), Universal Serial Bus (USB), and Ethernet. UAV software interfaces typically are:

- payload command and control,

- receipt of unmanned aircraft state data,

- sending payload status messages, and

- providing the payload data output stream for storage or downlink.

The data bus is the medium by which data are distributed across the avionics architecture. The data bus routes information over an electrical circuit from the data producers and consumers. The data are prioritized such that the most critical data have a high probability of getting through. U.S. military manned aircraft typically use a 1553 data bus [32] for avionics and payload systems.

7.4.6   C LANGUAGE

Typical high-level programming languages are:

- assembly,

- C, C++,

- Python, and

- JavaScript.

Writing a program in machine language would be tedious. Any program code written in a high-level language like C++ needs to be translated into machine code before it can be executed by a processor. This translation is done with a program aptly named Assembler.

C is an imperative procedural language [34]; it uses statements to specify actions. The most common statement is an expression statement, consisting of an expression to be evaluated, followed by a semicolon ";". By design, C provides constructs that map efficiently to typical machine instruction. There is a small, fixed number of keywords, including a full set of flow of control primitives such as "for," "if/else." "while." User-defined names are not distinguished from keywords. C has no "define" keyword; instead, a statement beginning with the name of a type is taken as a declaration. There is no "function"keyword; instead, a function is indicated by the parentheses of an argument list.

There are a large number of arithmetical and logical operators, such as "+" and "=". Procedures (subroutines not returning values) are a special case of function, with an untyped return type "void". Complex functionality such as I/O, string manipulation, and mathematical functions are consistently delegated to library routines. Comments may appear either between the delimiters "/*" and "*/" or following "//" until the end of the line.

The basic C source character set includes the characters such as lowercase and uppercase letters; decimal digits, and graphic characters. New software development tools improve development efficiency while reducing errors. Model-based design techniques allow development in the requirements and architectural level rather than working with lines of code from the start. Auto-coding generates software code based on higher-level design.

7.4.7   COMPILER

The name "compiler" is primarily used for programs that translate source code from a high-level programming language to a lower-level programming language (e.g., assembly language, machine code). Compilers and assemblers are used to convert both high-level and assembly language codes into a compact machine code for storage in the microcontroller's memory. Depending on the device, the program memory may be permanent, read-only memory that can only be programmed at the factory, or it may be field-alterable flash or erasable read-only memory. Programmable memory also reduces the lead time required for deployment of a new product.

7.4.8   ARDUPILOT

ArduPilot is an open source UAV platform [36], able to control autonomous multicopters, fixed-wing aircraft, traditional helicopters, ground rovers, and antenna trackers. ArduPilot is an award winning platform that won the 2012 and 2014 "UAV Outback Challenge" competitions. The system uses an Inertial Measurement Unit (IMU) using a combination of accelerometers, gyroscopes, and magnetometers. Today's ArduPilot is almost entirely C++ and has evolved to run on a range of hardware platforms and operating system.

7.4.9   DEBUGGING

The testing and debugging of the software is a very important and often time-consuming task is, which makes up a large portion of the overall development cycle. Testing is performed with the aim to check whether the programing meets its requirements. Detected deviations from the specification may result in debugging the program code (if its cause was an implementation error), but may even instigate a complete redesign of the software in case of a design flaw.

7.4.10   DESIGN PROCEDURE

In the preceding sections, basic fundamentals of a microcontroller and functions of its main components have been described. In this section, the design and development procedure of a microcontroller will be introduced. In general, the microcontroller development begins from the requirements analysis and ends with testing and maintenance. Figure 7.6 illustrates the microcontroller design/development procedure block diagram. The process has two few feedbacks to check if the requirements are met. The iteration continues until the design requirements are met. It should be noted that a flight simulation code including the controller design is developed in parallel. Thus, prior to the writing a code for the microcontroller, the controller design, and the flight simulation must be conducted. Thus, the flight controller must perform satisfactorily in a flight simulation, before it is included in the microcontroller code.

7.5   QUESTIONS

1. What is the major function of a microcontroller?

2. What are the primary functions of a microcontroller in a UAV autopilot?

3. What autopilot activities are performed via a microcontroller?

4. Name three differences between a microcontroller and a microprocessor.

5. Name at least three microcontroller designers/manufacturers/architectures/vendors.

6. What do A/D and D/A stand for?

7. Briefly describe the features of a microcontroller.

8. Describe Harvard memory Architecture.

9. Name at least five main modules of a microcontroller.

10. Name at least two high-level programming languages.

11. What is the function of a complier?

12. What are the differences between flight simulation software and the microcontroller code? Briefly describe.

13. Name at least three software to develop for a UAV.

14. Describe the procedure to develop the microcontroller code.

15. What are the ground payload software functionality?

16. Describe features of the ArduPilot.

17. Describe how a code (computer program) in Matlab (m-file) is converted and transferred to a microcontroller. In other words, write four main steps of the programming procedure.

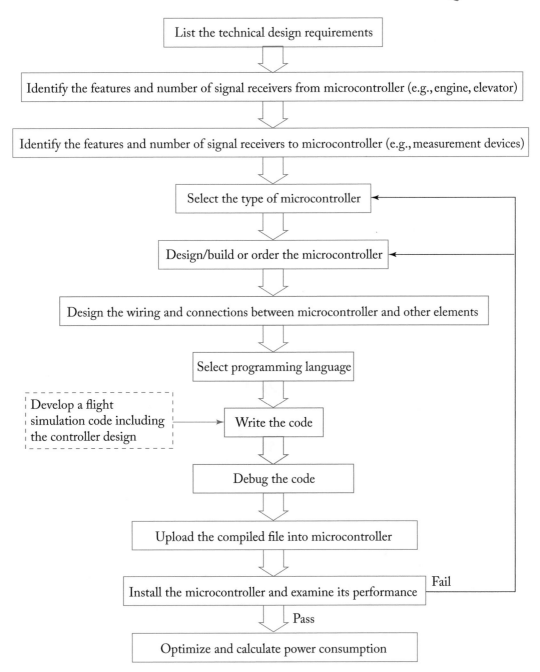

Figure 7.6: Microcontroller design/development procedure.

Part III

CHAPTER 8

Ground Control Station

8.1 INTRODUCTION

A significant element of a UAS system is the ground station, which is the man–machine interface with the unmanned aerial vehicle. Other alternative terms use in the literature are "Ground Control Station (GCS)", "Command and Control Station (CCS)", "Mission Control Station (MCS)", and "Mission Planning and Ground Control Station (MPGCS)." Depending upon range and type of mission (complexity of the UAV system), smaller UAVs are controlled via visual contact (manual real-time control), while the larger ones are equipped with a communication system (engage stored on-board flight plans). The human controller/pilot in the station will "talk" to the UAV via the communication system up-link in order to direct the flight or to update the flight plan. The ground station allows remote control of UAV, receives and processes navigation data from UAV via the data-link, supports flight planning, and communicates with the autopilot board on ground via interface. Direct operation of various payloads (e.g., camera, weapon) that the UAV carries may also be required. If data link is installed, ground station may support flight control without visual contact with UAV through software application. The flight plans are programmed by the operator; and uploaded through a serial connection between the UAV autopilot and the ground station. Three main functions for the ground station are mission tracking/planning, observing and piloting.

The aircraft will return information and images to the operators via the communications down-link (see Figure 8.1), either in real-time or on command. The information will usually include data from the payloads, status information on the vehicle's sub-systems, altitude and airspeed and position information. The launch and recovery of the vehicle may also be controlled from the station. A controller in the ground station has a place similar to a cockpit with displays, stick/wheel and buttons to control the UAV. The deflection to control surfaces and engine throttle are applied through communication signals. Conventional remote control is responsible for autopilot engagement and disengagement as well as allows manual/remote controlled flight operations during visual contact with the UAV. The UAV autopilot is controlled by human pilot in the ground station. This chapter will introduce features of ground station, main elements, and support equipment, and design requirement. Since a human is utilizing the station, the ergonomics is briefly reviewed.

In general, the operator(s) in the ground station have three main tasks, to: (1) control the UAV trajectory, (2) employ payloads, and (3) monitor sensors. If the UAV does not have the capability of full autonomous flight (e.g., Predator A), the controller in the ground station must engage with the vehicle in real time throughout the flight operation. However, for case of the full auton-omous UAVs, the operator(s) in the ground station will monitor the mission, and will act only if needed.

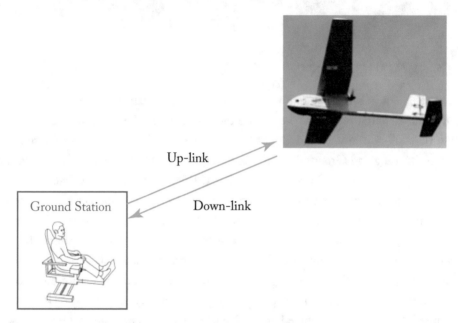

Figure 8.1: Ground station and the air vehicle (RQ-11 Raven UAV).

The typical operator's task includes: (1) control the UAV to orbit at a given altitude, radius, and speed around a to-be-designated terrestrial grid reference; (2) direct the UAV to hover over a selected area for surveillance; (3) descend at a controlled rate and land; and (4) descend to ground level to take an atmospheric sample and then to climb again to the operating altitude.

8.2   GCS SUBSYSTEMS

A GCS usually contain a number of sub-systems required to achieve its overall function. These will depend in detail on the type (civil versus military) and mission (e.g., range, endurance, and delivery) of the UAV. In general, the following sub-systems and elements may be structured into the GCS.

1. The UAV flight controls console. The operator will connect to the UAV autopilot (i.e., control system) to control the mission; either for manual real-time control of the UAV, or to select and engage stored on-board software.

2. The payload monitoring and control consol. The payload operator will monitor the status of the operators and control their operations (e.g., video camera).

3. Communication subsystem with UAV and other operators/commanders (encoders, transmitter and receiver). The communication with UAV has both uplink and downlink. This system includes the control for the status of communication elements. For example, raising or lowering radio antenna masts, steering them manually or automatically to obtain good transmission and reception of the radio waves, and changing frequencies as necessary. Section 8.5 provide more details.

4. Navigation displays for monitoring the status and flight-path of the UAV. The position of the UAV will be displayed automatically on a position display. Area navigation (RNAV) is a method of navigation that permits aircraft operation on any desired flight path within the coverage of station-referenced navigation aids:

5. display for recording equipment, and measuring payloads;

6. terrestrial map displays;

7. computers for data processing, calculations, mission planning, keeping the UAV flight programs, and housekeeping data storage;

8. equipment to provide comfort to operators (e.g., heater, air conditioner, dehumidifier);

9. support subsystem to provide electric/mechanical/hydraulic power to various equipment;

10. office equipment (e.g., table, seat); and

11. computational and communication software and computer programs.

8.3   HUMAN OPERATOR IN GROUND STATION

One of the most important aspects of operating a UAV is safety, which is heavily relying on external pilots. Operators in GCS directly control the UAV and its payloads with communication signals. The two most common personnel roles are air vehicle flight operator (or controller, or ground pilot) and mission payload operator. Other functions that may be performed in GCS are: (1) mission commander, (2) communication operator, (3) intelligent specialist, and (4) weather observer. Most UAV losses are attributed to operator errors; part of which is related to ineffective interface design. Some possible deficiencies of a weak GCS design are that the GCS: (1) creates excessive workload, (2) does not effectively present critical information, (3) does facilitate emergency procedures, and (4) has uncomfortable situation inside the GCS that generates operator stress.

The operator work environment should be comfortable; and the interface must be effective; otherwise there will be operative fatigue and probable UAV losses. Whenever an engineering device which deals with human is planned to be designed, the ergonomic standards need to be considered. Ergonomics (or human factors) is the science of designing user interaction with equipment and workplaces to fit the user. The field of human factors engineering uses scientific knowledge about human behavior in specifying the design and use of a human-machine system. The aim is to improve system efficiency by minimizing human error and optimize performance, comfort and safety. Proper ergonomic design is necessary to prevent repetitive strain injuries, which can develop over time and can lead to long-term disability. Salyendy [29] is a helpful resources in introduction, principles, fundamentals, and useful data for various aspects of human factors. In this section, a few important advices for optimizing the GCS are provided. For more details, Tso et al. [28] is recommended, which reviews human factors for command and control of unmanned air vehicles.

The operator within GCS must be able to access all relative information when needed. In the meantime, they should not be overwhelmed with the high volume of data. Moreover, the temperature and humidity at the GCS must be controlled to provide maximum comfort and effectiveness. In addition, the number of simultaneous activities should be less than or equal to a normal human capability. Thus, number of sticks, wheel, pedals, and displays should be limited. The seating environment and the seat/table relation must be designed based on the ergonomic fundamentals. The software/algorithm should be designed so that the interface is menu-driven, to allow the operator to find the necessary information in perhaps one to three mouse clicks. When there are more than one human operator inside the GCS, the tasks of each operator must be clearly defined.

Incorrect monitor positioning can cause neck and eye strain, and can lead to poor seat positioning, which creates pressure on the back. The top of the monitor should be positioned just above the eye level when seated. This is the best place for the "vision cone," the most immediate field of vision, which starts at the top at one's eye level and descends at a 30° angle. When monitors are too far away, people tend to lean forward to see well. This is increasingly true as people age, since vision almost inevitably declines over time. A rule of thumb: If one can extend his/her arm and just touch the screen with the fingertips, then he/she is in the right position.

To keep wrists and arms at an optimum position, reducing the risk of repetitive-motion injuries, the stick/yoke and switches should be at the same level as the elbows when seated. Since not everybody has a standard size, a simple fix is an adjustable seat. Sitting properly takes 20–30% of the pressure off the lower back. The seat should be between 17 in. and 19 in. deep, and it should have good lower-back support. The body should be positioned with the back against the seat and the hips open. If one finds himself/herself leaning forward to see the panel, he/she needs to move the seat forward.

Leg positioning contributes to the overall position in the seat, so make sure the legs are bent at about 90% angles at the knees. This helps alleviate pressure on the back. Movement is essential for circulation, however, so allow for subtle shifts in positioning and be sure to stand, stretch and walk a few steps at least every few hours. Feet should be firmly planted on the floor. If the seat positioning required for proper wrist alignment results in the feet not reaching the floor, use some types of footrest to support the feet, such that the height of the support keeps the knees at a right angle.

A human has various limitations (power, size, tolerance). Some noticeable ones are that a human: (1) needs rest, (2) is not very accurate, (3) may get sick, (4) cannot handle a load more than 12 "g", and (5) cannot work when pressure is beyond 0.75 – 1 atm. In contrast, a human is able to react to unexpected situations and does have an intuitive feel for an aircraft. Human factors must be considered in deriving design requirements.

8.4 TYPES OF GROUND STATIONS

Ground station range from a small handheld remote controller, to a simple laptop computer, to a mobile truck, to comprehensive fixed central command stations. The mobile stations are frequently located in all-terrain vehicles and are made for close- or medium-range systems. However, the central command stations are in fixed bases and are employed for MALE and HALE systems. In general, the control station do not necessarily need to be located on the ground; they may be positioned in a submarine, ship, another aircraft (airborne). In this section, a few types of ground stations are introduced.

8.4.1 HANDHELD CONTROLLER

Most small homebuilt UAVs and model airplanes are controlled via a simple handheld remote controller which weighs about 1-2 lb. The controller has a few sidesticks/levers/buttons (see Figure 8.2) with an antenna and uses radio/infrared signal to direct the UAV. These types of stations are utilized for visual and real time control. The operator will push/push stick/lever to deflect any control surface and to change the flight path.

8.4.2 PORTABLE GCS

When the necessary equipment for monitoring and controlling a UAV is slightly larger than the size of a handheld box, the solution is to adopt a larger portable GCS. This is true for mini to small UAVs. In some cases, both UAV and GCS can be back-packed and carried in a small suitcase. The size of such GCS is similar to the size of a laptop. Such GCS frequently incorporates the graphical user interface (GUI) and features a laptop, allowing the operator an easy access to key and frequently used features. A portable GCS requires a line-of-sight to allow the pilot to see the UAV and to send the command signals. Mission information is easily updated when the flight vehicle is re-tasked to support a changing environment. Figure 8.3 illustrates the mini UAV Desert Hawk III and its portable control station.

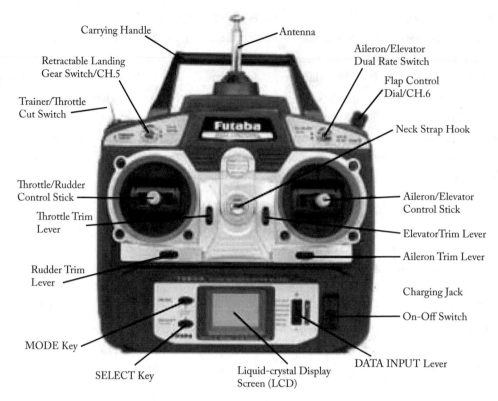

Figure 8.2: Handheld remote control of a small UAV (image courtesy of hooked-on-rc-airplanes.com).

8.4.3 PORTABLE GCS

When the necessary equipment for monitoring and controlling a UAV is slightly larger than the size of a handheld box, the solution is to adopt a larger portable GCS. This is true for mini to small UAVs. In some cases, both UAV and GCS can be back-packed and carried in a small suitcase. The size of such GCS is similar to the size of a laptop. Such GCS frequently incorporates the graphical user interface (GUI) and features a laptop, allowing the operator an easy access to key and frequently used features. A portable GCS requires a line-of-sight to allow the pilot to see the UAV and to send the command signals. Mission information is easily updated when the flight vehicle is re-tasked to support a changing environment. Figure 8.3 illustrates the mini UAV Desert Hawk III and its portable control station.

8.4.4 MOBILE TRUCK

The GCS for close-range UAVs have a number of equipment and devices which do not fit in a portable box. The volume and weight of such equipment are such that they can be setup in back of a truck. A few advantages of a truck are: (1) the truck can transport the UAV, GCS, and the human operator to any designated area, (2) the truck can provide the GCS with electric power, (3) the back of a truck can be converted to a comfortable place for the UAV controller, and (4) the truck will keep the human operators in the loop. A truck easily meets the mobility requirement of the UAV system. Thus, the GCS for close-range UAVs are usually mobile and housed within an "all-terrain" vehicle. A pneumatically raised steerable radio mast is at the rear of the vehicle which carries antennae for communication with the UAV.

Figure 8.3: Desert Hawk III mini UAV and its portable control station(image courtesy of defense-update.com).

The air vehicle of these systems usually will be either ramp-launched or VTOL. The requirements and capabilities of both systems will be generally similar, except for the specifics of the control during launch and recovery. Although unmanned aircraft system implies that no human operator is involved, but these systems might require the manpower equivalent to manned aircraft. There is no sensation of speed, attitude, or g's in any maneuver. Each operator will be assigned an alternative assignment task; plus a crew member as the commander who is responsible for the inter-system communication. There is usually a flexibility of tasking and easy cooperation between crew members.

The ground vehicle is usually fitted with chassis-mounted jacks or stays which are lowered once the truck is on-site. These are necessary to stabilize the truck to prevent it from rocking under the influence of wind or the operators moving around within the vehicle. The GCS also incorporates electrical power generation and air-conditioning which is required not only for the comfort of the crew, but for the climatic control of electric equipment (e.g., monitors, computers). These equipment also have their own controls and monitoring systems provided in a separate rack. There will be a longer duty time of the operators, additional crew members may be required, especially for the more complex payloads.

There are a number of monitors and displays. For instance, a monitor carries a video image from a TV camera. Beneath the monitors are the control decks, over which are the control sticks to direct operator control of the UAV in flight. The keypad controls for inputting mission data, software updates, or way-points to the UAV in located on the control deck. Controls are provided for starting the aircraft, selecting pre-flight test data, and for activating the aircraft launch. There will be displays showing the payload status and data (imagery and/or other types of data) with recording media. Thus, a major GCS section is the sub-system to recognize the type of payload installed in the UAV, and to control the operation of the payload. A display may also be included to show communication status. The display screen should be large enough, so that the operator's eye is able to take advantage of the full resolution of UAV sensors.

The operators in the GCS are not visualizing the UAV that they are controlling. All they see is an icon appears on a moving map display, and the payload imagery which is displayed on a monitor. The UAV could be aesthetically pleasing, large or small, but the icon on the map makes no distinction. Figures 8.4a-b illustrate the RQ-7A Shadow 200, its GCS, and inside of GCS (e.g., an operator, a few displays can be seen).

When a close range UAV is launched from sea, the control station will be naturally located in a ship. The Insitu ScanEagle UAV (with 20+ hr endurance; Figure 2.5) is an example where the GCS is placed onboard a navy ship. A ship, similar to a truck, is capable of providing comfort for operators, transporting GCS and UAV, and generating electric energy for GCS. However, there is a possibly that a GCS be airborne; in another aircraft (i.e., parent' aircraft).

(a) RQ-7A Shadow 200 and its GCS

(b) The inside of the RQ-7A Shadow 200 GCS

Figure 8.4: RQ-7A Shadow 200.

8.4.5 CENTRAL COMMAND STATION

For the case of a UAV (e.g., Global Hawk and Predator) with a long range (more than 10,000 km), and a very long endurance (more than 10 hr), the GCS must be comprehensive and be located in a fixed location. There are generally additional missions and associated payloads beyond the common types; only a fixed-base GCS may accommodate most potential types with a single GCS. A fixed GCS will provide more spacious accommodation, so operators are provided comfort for a longer duty time. A fixed base GCS is a building with complete office and life equipment. It will accommodate several flight crew members, payload operators, and commanders. A challenging part in the design of a very long endurance UAV is the design of its fixed-based GCS.

A long-range UAV will usually takeoff horizontally on wheels along a runway. Provision has to be made to control the aircraft during its take-off and lift-off. This and subsequent recovery is accomplished under direct operator control, usually with the aircraft in direct view of the operator. Additional crew members may be required, especially for the more complex payloads carried. Moreover, there are more crew involved the control of the UAV and conducting the entire mission. Most modern UAVs have a mission payload operator workstation. This position provides an operator with the ability to command payload functions and utilize the payload data. If the UAS is equipped with an EO (electro-optic)/IR (Infra-Red) ball or perhaps a SAR, these payloads are controlled and managed from the baseline workstation.

Figure 8.5: Global Hawk Operations Center at NASA Armstrong.

A long-ranging radio equipment are employed for wider networking. Crews will operate in shifts during the long flight-times of the UAV. Provision will therefore be made for handing over responsibility for the mission. If the aircraft carries armament, a further crew member, the Weapons Systems Operator, is required to select, monitor, release and guide the weapons onto the target. UA Systems such as Predator and Global Hawk have the option to launch the UAV from a GCS on an airfield relatively close to the target area, but, after launch, be controlled from a command center which may be a few thousand kilometers away. The Global Hawk system employs two ground-based GCSs. They are the launch and recovery element, known as the satellite control station, and a mission control element.

The following items will be held in a fixed GCS: operating manual, maintenance manual, consumable items, repair tools, spare parts, and special test equipment. Figure 8.5 illustrates the Global Hawk (Figure 1.1) Operations Center at NASA Armstrong. In the fall of 2010, flight crew and scientists occupied the GCS for the Genesis and Rapid Intensification Processes hurricane study.

8.5   COMMUNICATION SYSTEM

An important subsystem of an unmanned aircraft is the communication system which provides primarily an uplink and downlink. The principal issues of communication technologies are flexibility, adaptability, security, and cognitive controllability of the bandwidth, frequency, and information/data flows. A UAV data link typically consists of an RF transmitter and a receiver, an antenna, and a modem to ink these parts with the sensors. Data link spoofing, hijacking, and jamming are major security issue facing communication system.

The up-link transmits command and control from the operators to the UAV (or UAVs in multiple operations), and the down-link returns payload data, and the flight status and images from the UAV to the GCS and to any other satellite receiving stations. The UAV status data (e.g., location, airspeed, attitudes, and weather conditions) is often known as housekeeping data. Loss of communication during flight operations will likely result in the UAV crash. The ground elements the communication link between the GCS and UAV includes: (1) encoders, (2) transmitter, (3) receiver, and (4) controls for their operation. The range of communication system is a function of the heights of the radio antenna and air vehicle, these two determine the availability of LOS. As the UAV altitude increases, the LOS range will be increased too. An important application of communication is the avoidance of mid-air collisions between UAV and other aircraft.

The communication between the GCS and aircraft and between the aircraft and GCS may be achieved by two different media: the radio or the laser beam. The communication by radio between the UAV and the GCS can be direct or via satellites or other means of radio relay. Typical usable radio frequencies for communication range from 3 HZ to 300 GHZ (below the infrared spectrum of electromagnetic waves). Frequencies in the range 3 Hz (extremely low frequency, ELF) to 3 GHz (ultra-high frequency, UHF) are effective radio frequencies as they are refracted in the lower atmosphere to curve to some degree. The selection an operating frequency requires a compromise. A lower frequency offers a better and more reliable propagation, while having reduced data-rate ability. In contrast, a higher frequency is capable of carrying high data rates, but requires direct LOS and higher power to propagate the signal. For instance, a high-resolution TV camera will produce a data rate of order 75 megabytes per second.

Telemetry is the automatic measurement and wireless transmission of data from UAV to a ground station. In this process, sensors at the UAV measure electrical data and flight data (such as altitude, airspeed, and angular rates). These measurements are converted to specific electrical voltages. A multiplexer combines the voltages, along with timing data, into a single data stream for transmission to a ground station using a radio signal. Upon reception, the data stream is separated into its original components and the data is displayed and processed according to operator specifications. It can also be used to alert GCS when battery levels are too low or when a fault occurs.

The most applicable antennae for UAVs are: (1) vertical antenna, (2) Yagi-Uda antenna, (3) parabolic dish antenna, and (4) phased array rectangular micro-strip or patch antenna. Figure 8.6 illustrates a NASA Altair satellite antenna. It is always desirable to have the radio antennae mounted on the GCS to minimize conductor lengths. Meanwhile, it is necessary to situate the antennae in a position to obtain good transmission and reception of the radio waves. In a hilly terrain area, this may mean on a crest. For some military operations, unless the GCS is small and readily concealed, its position may become obvious to hostile forces and so vulnerable. A solution would be to mount the antennae on a small, mobile, and less-detectable platform which can be located in a suitable position.

General Atomics RQ-1 Predator UAV (Figure 5.5) was the first UAV to be controlled via satellite communications, and the first UAV to provide voice radio communications. Large modern UAVs (e.g., MQ-9 Reaper (Figure 4.10), and NASA Altair (Figure 9.1)) are equipped with satellite antenna. The Ground Control Station for RQ-1A Predator uses C- and Ku-Band datalinks for Line-Of-Sight and non-LOS communication with the UAV, respectively. The range of the non-LOS link, and therefore the effective operational radius of the aircraft, is about 740 km. Non-LOS communications in the first three Predator systems was handled via a Trojan SPIRIT SATCOM link.

Figure 8.6: **NASA Altair satellite antenna.**

There are two techniques to track a UAV is by radio communication: (1) fit the UAV with a transponder which will receive, amplify and return a signal from the control station, and (2) UAV down-link should transmit a suitable pulsed signal. The successful operation of the communication system depends upon the integration of the various components of the system to supply adequate energy to achieve the required range. Four contributing factors are: (1) transmitter power output, (2) receiver sensitivity, (3) antenna gain, and (4) path loss.

These communications for UAV are done primarily through the use of RF applications, usually, satellite communication links in UAV are used LOS mode. The most common frequency bands of this type of links are Ku, K, S, L, C, and X bands. The Ku band has been historically used for high speed links. Due to its short wavelengths and high frequency, this band suffers from more propagation losses. Yet it is also able to trespass most obstacles thus conveying great deals of data. K band possesses a large frequency range which conveys large amounts of data. However, it requires powerful transmitters and it is sensitive to environmental interferences. The S and L bands do not allow data links with transmission speeds above 500 kbps. Their large wavelength signals are able to penetrate into terrestrial infrastructures and the transmitter require less power than in K band. The C band requires a relatively large transmission and reception antenna. The X band has been reserved for military application. Table 8.1 shows the commonly used frequency bands [39] for communication systems.

Table 8.1: Commonly used frequency bands in communication		
No	Band	Frequency
1	HF	3-30 MHz
2	VHF	30-300 MHz
3	UHF	300-1000 MHz
4	L	1-2 GHz
5	S	2-4 GHz
6	C	4-8 GHz
7	X	8-12 GHz
8	Ku	12-18 GHz
9	K	18-26.5 GHz
10	Ka	26.5-40 GHz

Figure 8.7 illustrates the GCS Command, Control, and Communications (C3) model. Technologies and operating procedures related to C3 of UAVs are divided into two categories: (1) RF line of sight, and (2) RF beyond line of sight. Under each category UAV technical issues may be divided into two approaches: (1) Command and (2) Control and Air Traffic Control (ATC). Under Command and Control, both full autonomy, and remote control are applicable.

Designing UAV wireless datalink is much more challenging than other wireless links. The key challenges are long distance, high speed, and spectrum. New data links need to be developed for UAVs and for commercial manned flight, since they will share the same air space and would need to be aware of each other's presence. The key challenges in the design of UAV communication system are the large distances that they need to cover and the high speed of aircrafts. These along with the limited availability of radio frequency spectrum affect the performance of the data link.

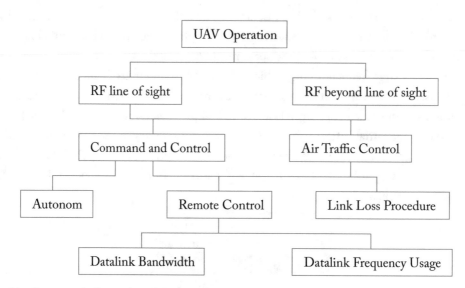

Figure 8.7: Command, Control, and Communications (C3) model.

8.6   DESIGN CONSIDERATIONS

In the preceding sections, various types of GCS have been described. In designing GCS for a UAV, one must select the type of GSC, select the measurement devices, number of operators, and then conduct some calculations and analysis. In general, the primary criteria for the design of GCS are as follows: (1) manufacturing technology, (2) required accuracy, (3) mission, (4) weather, (5) reliability, (6) life-cycle cost, (7) UAV configuration, (8) human factors, (9) maintainability, (10) endurance, (11) communication system, (12) weight, and (13) level of control.

Figure 8.8 illustrates a flowchart that represents the GCS design process with a number of feedbacks. In general, the design process begins with a trade-off study to establish a clear line between cost and performance (i.e., accuracy) requirements; and ends with optimization. The designer must decide about two items—select type of GCS and select the desired equipment. After conducting the calculation process, it must be checked to make sure that the design requirements are met. A very crucial part of the design process is to integrate the GCS with the air vehicle. If all equipment/displays are selected/purchased, the integration process must be still conducted. The comfort level and effectiveness of the operators (ergonomic indicators) must be determined in order to make sure that the mishap and error is minimal.

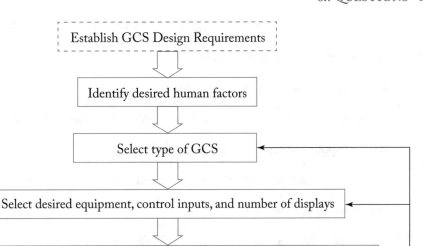

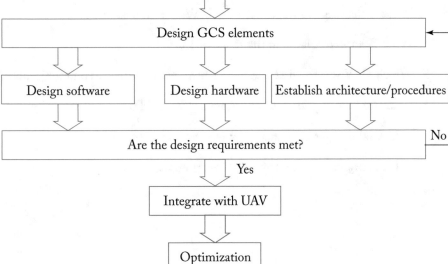

Figure 8.8: GCS design process

8.7 QUESTIONS

1. What are the alternative terms equivalent to the GCS?

2. List GCS Subsystem in brief.

3. What is ergonomics?

4. What are a few outputs to operator when the GCS is not optimally designed?

5. Name four main types of GCS.

6. At what situation does the designer need to select a fixed comprehensive GCS?

7. What are the features of a mobile truck as the GCS?

8. Describe the features of a portable GCS.

9. What are the primary functions of the UAV communication system?

10. List applicable types of antenna to UAVs.

11. What are the four main elements of the communication system in the GCS?

12. What is the housekeeping data?

13. What is the range of frequencies for UAV communication?

14. What factors contribute in the range of LOS?

15. Which UAV was the first one to be controlled via satellite communications?

16. Compare human pilot with an autopilot by providing:

   a. three advantages of human pilot over an autopilot, and

   b. three advantages of an autopilot over a human pilot.

CHAPTER 9

Launch and Recovery Systems

9.1 INTRODUCTION

All unmanned aircraft must initially took-off/be-launched to become airborne. Moreover, at the end of the flight mission, they must land on an airfield, or be recovered. Launch involves transitioning the UAV from a nonflying state to a flying state. In the case of a conventional runway launch, this can be considered takeoff. Various conventional and unconventional launch and recovery techniques have been applied to manned aircraft for over a century. The range of options is greater for unmanned aviation. These techniques are largely enabled by the exclusion of pilot physical constraints and the inclusion of much lower UA weights.

Typical launch methods include: (1) rail launchers, (2) rocket launch, (3) air launch, (4) hand launch, (5) tensioned line launch (i.e., catapult), (6) gun launch, and (7) ground-vehicle launch. Currently, the following recovery methods are employed: (1) skid and belly recovery, (2) net recovery, (3) cable-assisted recovery, and (4) parachutes.

The vast majority of large fixed-wing UAVs employ conventional launch and recovery techniques, i.e., conventional takeoff and landing. Conventional launch and recovery requires a length of horizontal flat surface (runway); and a further distance before the nearest obstacle of maximum specified height. Of all the methods for UAV launch and recovery, the VTOL one is an accurate, operationally convenient, and gentle. VTOL are more appropriate for very small UAVs.

When a UAV has a conventional takeoff and landing, the Launch and recovery systems is converted to the design of landing gear. The approach for designing the landing-gear for a tactical-to-large UAVs, such as the AAI RQ-7 Shadow 200 or the Northrop Grumman RQ-4 Global Hawk, is no different than for a manned aircraft. The high-altitude long-endurance UAV, Global Hawk (see Figure 4.1) has a high speed required for lift-off (about 100 knot). It is powered by a single turbo-fan engine whose thrust is just over 40 kN and requires a ground run of over 600 m. The landing gear design is out of scope of this book, landing gear design process is presented in Sadraey [37]. In this chapter, the launch of micro-to-small UAVs are addressed, which particularly using concepts that avoids the need for a runway.

9.2   FUNDAMENTALS OF LAUNCH

The analysis of launch and design of launcher require an understanding of the concept of launch, and the application of governing equations. In this section, parameters, contributing forces, and governing equations of the launch operation are presented.

The relationship between acceleration, launch speed, and launcher length has significant implications for launcher design. The integral form of the equation is recommended because launch force is rarely constant in practice. The peak loads are often significantly higher than the average loads. From theory of dynamics, when a moving object with an initial velocity of V_1 accelerates to a new velocity of V_2, the distance (x) covered is governed by the following equation:

$$V_2^2 - V_1^2 = 2ax \tag{9.1}$$

where a is the linear acceleration. For a UAV launcher, the initial velocity is zero, and the distance traveled is the length of launcher (L_L). So, the governing equation is expressed as:

$$V_L^2 = 2aL_L \tag{9.2}$$

In addition, in order for a launcher to create such acceleration, it must provide a sufficient launch force (FL). From Newton's second law,

$$F_L - F_f - W\cos\theta = ma \tag{9.3}$$

where θ is the launcher angle (see Figure 9.1), F_f is the friction between launcher rails and the UAV (indeed, two metals), and W is the weight of UAV ($W = mg$). The friction force is proportional to the normal force to the launcher:

$$F_f = \mu N = \mu W \cos\theta \tag{9.4}$$

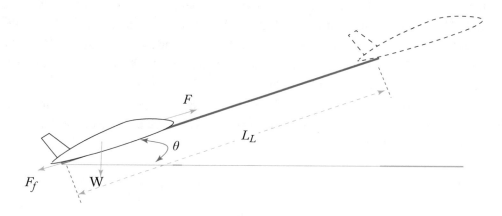

Figure 9.1: Contributing forces on a launcher.

The friction coefficient (μ) between a UAV and the launcher rails is typically about 0.05–0.06. When the friction force is substituted from Equation (9.3) into Equation (9.4), we obtain:

$$F_L - \mu W \cos\theta - W \cos\theta = ma \rightarrow F_L = W \cos\theta \, (1 + \mu) + ma \qquad (9.5)$$

This is the force that the launcher must provide. The source of power to provide launcher force may be of pneumatic, or hydraulic, or a spring. As the launch angle is increased, the greater launcher force is required. A compromise is needed to select the launcher length, since a long launcher requires less force, but makes the launcher heavier and needs a larger volume for depot. For more accurate calculation, you may include UAV lift and drag forces.

The UAV speed at the end of launcher is often about 10–30% greater than the stall speed. The UAV stall speed is a key parameter governing the energy of launch event. It is desirable to reduce the launch energy in order to lessen the UAV loads and the footprint of any supporting apparatus.

$$V_s = \sqrt{\frac{2W}{\rho S C_{Lmax}}} \qquad (9.6)$$

where S is the wing area, ρ is the air density (at sea level, 1.225 kg/m³), and C_{Lmax} is the maximum lift coefficient (about 1.2–1.6). The equations in this section are employed to calculate the launcher length and required launcher force. The total launcher length should not exceed practical transportability and storage dimensions, but the rail may consist of multiple segments. Many governing equations of a normal takeoff operation is applicable to launch. Sadraey [38] provides details of takeoff operation and introduces its governing equations and contributing parameters.

Example 9.1: A launcher is designed to launch a UAV with a mass of 500 kg. The engine thrust is assumed to be constant and equal to 1,000 N. The launch angle is 20°, the launcher length is 3 m, and the friction coefficient is 0.06. The UAV is required to reach a speed of 30 knot at the end of launcher rail. What force is needed to push the UAV over the launcher?

Solution:

$$V_L^2 = 2aL_L \rightarrow a = \frac{V_L^2}{2L_L} = \frac{30 \times 0.514^2}{2 \times 3} = 39.6 \, \frac{m}{s^2} \qquad (9.2)$$

$$F_L = W \cos\theta \, (1 + \mu) + ma = 500 \times 9.81 \times \cos(20) \, [1 + 0.06] + 500 \times 39.63 = 25,300 \, N \qquad (9.5)$$

9.3 LAUNCHER EQUIPMENT

UAVs can be launched from moving ground vehicles. A launcher eliminates the need for conventional landing gear for the purposes of takeoff. The launcher is a mechanical devise to accelerate a fixed-wing UAV to a minimum controllable airspeed before releasing it from launcher. The ground vehicle provides the acceleration to flight speed, at which time the UAV is released. This launch technology predates conventional landing gear on flying aircraft. UAVs which do not have a vertical takeoff capability, a launcher will be required. In the design of a launcher, a few parameters such as launcher length, launcher weight, launch angle, and the required force and power must be calculated. Moreover, the type of source of launcher power (e.g., mechanical, spring, pneumatic and rocket) needs to be decided. The major elements of a launcher are: (1) two parallel rails, (2) pitch angle adjustment mechanism, (3) mobility wheels, (4) transportation truck, (5) power source, and (6) push mechanism.

Rail launchers provide the UAV with a stabilizing track and launch energy to take the UAV from rest to the airborne. The UAV is secured to a shuttle that travels along the rail via guides. It usually takes the form of a ramp along which the UAV is accelerated on a trolley, propelled by a system of rubber bungees, by compressed air or by rocket, until the aircraft has reached an airspeed at which it can sustain airborne flight. The launch method must provide safe physical separation between the UAV and the launcher; and other hazards throughout the launch operation. Most rail launchers with pneumatic/hydraulic power unit are capable of launching UAVs with a weight range of 500–1,000 lb. Typical launch speed is about 50–70 knots.

Figure 9.2: UAV Shadow and its launcher.

Pneumatic pistons are the most common tool to providing the launch energy. A compressor is required to pressurize air in the accumulator. Launchers must have a reacting mass or be fixed to the ground for stability. Figure 9.2 illustrates the Shadow UAV and its launcher. The Insitu Aerosonde was car launched for the famous transatlantic flight, enabling a lightweight airframe without gear.

To minimize the length of runway/launcher, rocket engine/booster can be employed to launch a UAV. The rocket's high thrust-to-weight ratio and modularity can enable simple installation. Rocket launch is appropriate when no runway is available and other techniques such as pneumatic rail launchers have excessive footprint. The primary forces acting on the UAV are the rocket engine thrust the UAV weight, and the normal propulsion system thrust. Hence, the rocket engine force is addition to the conventional engine thrust. It may be used on both runway and launcher. After the launch, the rocket engine may be dropped.

9.4   RECOVERY TECHNIQUES

Recovery systems eliminate the complexities of runway landing. Recovery is defined as transitioning a UAV from a flying state to a nonflying state. In the case of a conventional runway recovery, this phase is referred to as landing. Landing involves the return of the UAV to make a controlled touch-down onto its landing gear at the threshold of runway, deceleration along the runway, followed by the UAV taxiing or being towed back to its hangar. UAV recovery can be more challenging than launch, and touch-down at the correct position and airspeed require considerable judgement. For long-range UAV such as Global Hawk, a form of approach control with guidance, along a radio beam may be used. Initial positioning onto the radio beam will be performed with the help of GPS.

Recovery systems eliminate a need for any runway. Recovery requires that the UAV horizontal and vertical velocity components become equal to that of the recovery platform (in most cases, it means a stationary status). Recovery of the UAV requires a safe landing, as well as the return of the UAV to its base or hangar. For UAVs weighing more than approximately 500 lb, the absorption of recovery energy by any means other than landing gear is challenging and almost impractical. A few practical recovery techniques (other than conventional landing) include: (1) parachutes, (2) arresters, (3) nets, and (4) suspended cables with hooks.

A few techniques for recovery of a catapult launched UAV are: (1) guided flight into a catchment net, (2) skid or belly landing, (3) guided flight onto an arresting pole and suspended cables, and (4) deployment of a parachute—in flight—to reduce the ground impact. Suspended cables seizes both the horizontal and vertical motion of the UAV relative to the recovery platform. The arresting cables on the horizontal platform only arrest horizontal motion and act in conjunction with a conventional runway landing. The dynamics have many similarities to the net recovery technique. The Insitu ScanEagle UAV was purposely designed to use two technique, the net and the hook (Figure 9.3). The vertical rope requires an airframe-mounted hook with a horizontal extension. The Insitu ScanEagle UAV with a maximum takeoff mass of 22 kg, a wing span of 3.1 m, is equipped with a 2-stroke piston engine (1.5 hp) and has a maximum speed of 80 knot, and a service ceiling of 19,500 ft.

Skid and belly landing is a recovery technique where the UAV contacts the ground directly without the use of conventional landing gear. Either a skid device or the UAV fuselage is the interface to the ground, providing both landing shock absorption and friction to slow the UAV to a rest. Skids or body ground interface points must be designed to contend with the friction of scraping the ground.

Net recovery is a recovery technique where the UAV flies into a net, and its motion relative to the net (i.e., platform) is arrested. This technique is commonly used for larger UAVs that are too heavy for simple belly landings. Unmanned aircraft heavier than 500 lb typically avoid this technique in favor of conventional landing or parachute recovery.

(a) Net

(b) Skyhook

Figure 9.3: ScanEagle recovery.

9.5   RECOVERY FUNDAMENTALS

In Section 9.4, various ground recovery techniques have been introduced. In this section, the fundamental principles of two recovery techniques are briefly reviewed.

9.5.1   PARACHUTE

One approach for recovery for UAVs without a vertical flight capability, is using a parachute; which will allow a wheeled or skid-borne landing onto terrain. A parachute is installed within the UAV, carried from takeoff, and is deployed at a suitable altitude over the landing zone. As the parachute gets bigger, the impact to the ground will be slower. However, a large parachute is heavy and adds to the weight of UAV. So, a compromise needs to be made in selecting the parachute size. The design of parachute is based on the drag concept and terminal velocity. The landing/impact velocity will be equal to the terminal velocity (V_t). The terminal velocity relation is obtained by equating the drag and weight:

$$V_t = \sqrt{\frac{2W}{\rho S C_D}} \tag{9.7}$$

where W is the weight of UAV (including the parachute), S denotes the parachute projected area, and C_D is the parachute drag coefficient. Typical values for parachute drag coefficient are about 1.5–1. The landing velocity of the UAV relative to the platform is a key parameter governing the energy of recovery event. It is desirable to minimize the recovery energy in order to reduce the UAV impact force and the footprint of any supporting apparatus. Figure 9.4 illustrates the SkyLite UAV recovery using a parachute.

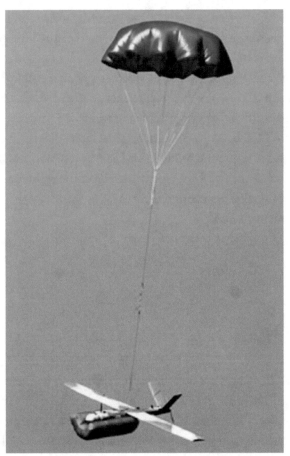

Figure 9.4: SkyLite UAV recovery (image courtesy of defense-update.com).

Example 9.2: A small UAV with a mass of 100 kg is required to land on the ground with the help of a parachute with an impact velocity of 0.5 m/s. What parachute projected area is needed for landing on sea level altitude? What is its radius?

Solution:

At sea level, the air density is 1.225 kg/m^3. The parachute projected area is determined by using Equation (9.7). For parachute drag coefficient, we take the average of 1.75.

$$V_t = \sqrt{\frac{2W}{\rho S C_D}} \rightarrow S = \frac{2W}{\rho C_D V_t^2} = \frac{2\times100\times9.81}{1.225 \times 1.75 \times 0.5^2} = 4100 \text{ m}^2 \tag{9.7}$$

$$S = \pi R^2 \rightarrow R = \sqrt{\frac{S}{\pi}} = \sqrt{\frac{4100}{\pi}} = 36.1 \ m$$

9.5.2   IMPACT RECOVERY

Other alternative forms of recovery equipment are arresters, a nets, and suspended cables with hooks. The recovery system design for these cases is based on the impulse-momentum theorem. Two techniques which are mainly utilized for micro to mini-UAVs are net and skyhook. In both cases, a large net or, alternatively, a carousel apparatus, or a hanging hook is employed to which the UAV is flown and caught. A means of absorbing the impact energy is needed, usually comprising airbags or replaceable frangible material. The arresting technique which the impulse-momentum equation is applicable is for larger UAVs. The recovery system reaction forces that bring the UAV to rest initially should be quickly disengaged to prevent sending the UAV in rearward and upward motions.

An impulse is equal to the net force (F) on the object times the time period (t) over which this force is applied. The impulse-momentum theorem stats that the impulse in an impact is equal to the change in linear momentum.

$$Ft = m \ (V_2 - V_1) \tag{9.8}$$

Since the final velocity is zero:

$$Ft = mV_2 \tag{9.9}$$

The energy of this impact (i.e., kinetic energy) must be absorbed by net, hook, or arrester without any harm to the UAV. The size of the net is a function of the UAV frontal area, as well as the desired reliability. The larger the net, the more reliable will be the recovery system. The net should be flexible so that the UAV get caught in a safe way. The flexibility will provide a longer time (t) for the UAV impact. The more energy absorbed for recovery, the greater the safety risk. The impact velocity and altitude for a given tension are controlled by the hook location.

9.6   AIR LAUNCH

Other than ground vehicle launch, there are a number of techniques that a UAV can be launched. One launch technique is to employ another aircraft (manned or unmanned) to release a UAV. Air launch involves releasing the UAV from a host or mother platform (i.e., another aircraft) at altitude and with high initial airspeed relative to ground. The unmanned aircraft can be released from a host platform's internal bay, under wing, under the fuselage, or from a cargo door. The mother aircraft have a stations that UAV is attached to. Typically, a launch rail will attach to a station, and that launch rail is what carries the UAV. The launch method depends greatly on the mission and characteristics of both the mother aircraft and UAV. Usually, the UAV is dropped from under the host platform. The design of the UAV air launch system is challenging, and involve a lot of calculation, and may generate a risk to the mother aircraft. This technique is only used when the UAV does not have a long range to reach the target area. So an aircraft carries the UAV to an area close to the target zone, and then release/launch it.

9.7   HAND LAUNCH

A very cheap and easy launch technique for a very small and light UAV is the human operator. This approach achieve runway independence. Very small unmanned aircraft can employ a hand-launch technique, which eliminates the need for any launch equipment. An overhead launch is the most common form for launching very small UAVs. Indeed, the human becomes the launch equipment. The individual launching the UAV is able to provide initial velocity and orientation, to enable controlled initial flight conditions. The operator must have sufficient hand force to launch the UAV. Moreover, he/she may run while launching to increase the initial velocity. After the hand launch, the UAV will use its engine force to continue the flight. Hand launch will drive the UAV configuration; one example is eliminating a need for the landing gear. Figure 9.5 shows the launch of the small UAV Pointer by hand.

A typical military male operator may provide a hand force of up to 100 N. Moreover, the length of an extended hand is about 70–90 cm. By application of Newton's second law and Dynamics governing equations, one can determine the maximum mass (m) of a UAV that can be launched by hand.

$$F = ma \tag{9.10}$$

Moreover, the operator may run during the launch period in order to increase the initial speed. When a hand launch system is selected for a UAV, no design activity is required for the launch system. However, an analysis should be conducted to make sure that the operator can provide a safe initial velocity.

Figure 9.5: Pointer launch.

Example 9.3

A very small UAV has a mass of 1.5 kg. A human operator with a maximum force of 90 N is arranged to launch this UAV. If his hand extends 70 cm, determine the release speed.

Solution:

The Newton's second law yields the linear acceleration:

$$F = ma \rightarrow a = \frac{F}{m} = \frac{90}{1.5} = 60 \ \frac{m}{s^2} \tag{9.10}$$

Release speed:

$$V_2^2 - V_1^2 = 2ax \rightarrow V_2 = \sqrt{2ax} = \sqrt{2 \times 60 \times 0.7} \rightarrow V_2 = 9 \ \frac{m}{s} \tag{9.1}$$

9.8 LAUNCH AND RECOVERY SYSTEMS DESIGN

Design of launch and recovery systems is a challenging task, and requires a large amount of calculations. Significant design requirements for the launch and recovery system of a UAV are: performance, cost, stability, reliability, safety, manufacturing issues, weight, size, marketability, and handling requirements. Figure 9.6 presents the general design process of launch and recovery systems. The design has an iterative nature and begins with the design requirements.

Two major sections of launch system design and recovery system design are designed in parallel. At any stage of the design procedure, the output is checked with the design requirements, and a feedback is taken. The process is repeated until the requirements are met. The primary criteria for the design of an autopilot are as follows: (1) manufacturing technology, (2) required accuracy, (3) stability requirements, (4) structural stiffness, (5) load factor, (6) size, (7) safety, (8) reliability, (9) cost, (10) UAV configuration, (11) maintainability, (12) weight, (13) aerodynamic considerations, (14) mobility/transportation, and (15) operational requirements.

Since there are various techniques for launch and recovery systems, no details are provided for the design procedure. Each approach has a unique group of equipment, and entails a specific design procedure.

Butler and Loney [55] presents the design, development, and testing of a recovery system for the emergency flight termination of an 1,800-lb air vehicle, the General Atomics Aeronautical Systems (GA) Predator Medium Altitude Endurance Unmanned Air Vehicle (MAE-UAV).

9.9 QUESTIONS

1. List main design requirements for the launch and recovery system of a UAV.

2. In the design of a launcher, what parameters must be determined?

3. What are the major elements of a launcher?

4. What is the function of UAV recovery system?

5. Name a few practical techniques in the UAV recovery.

6. What technique is employed in the recovery of (a) SkyLite UAV, (b) Insitu Scan Eagle, (c) Global Hawk?

7. Describe the procedure of a UAV hand launch.

8. What major elements are required for the air launch of a UAV?

9. What technique is used in launching Pointer UAV?

10. What is a typical impact speed of a UAV when it is recovered by a parachute?

11. What is the maximum takeoff mass and wing span of ScanEagle UAV?

12. Describe the recovery system of ScanEagle UAV.

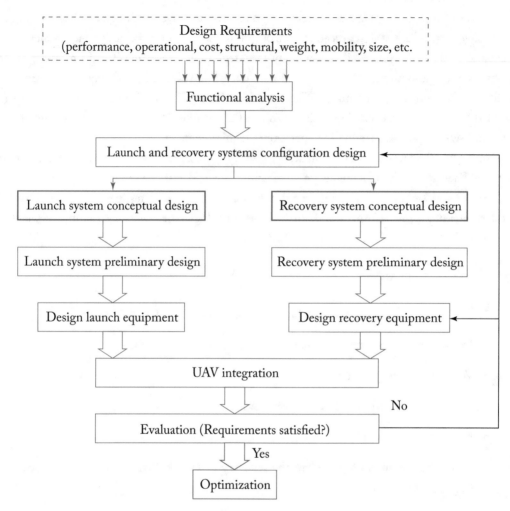

Figure 9.6: Launch and recovery systems design process.

9.10   PROBLEMS

1. A launcher is designed to launch a UAV with a mass of 300 kg. The engine thrust is assumed to be constant and equal to 500 N. The launch angle is 25°, the launcher length is 2.6 m, and the friction coefficient is 0.05. The UAV is required to reach a speed of 20 knot at the end of launcher rail. What force is needed to push the UAV over the launcher?

2. A launcher is designed to launch a UAV with a mass of 200 kg. The engine thrust is assumed to be constant and equal to 300 N. The launch angle is 30°, the launcher length is 2.2 m, and the friction coefficient is 0.055. The UAV is required to reach a speed of 15 knot at the end of launcher rail. What force is needed to push the UAV over the launcher?

3. A small UAV with a mass of 70 kg is required to land on the ground with the help of a parachute with an impact velocity of 0.3 m/sec. What parachute projected area is needed for landing on sea level altitude?

4. A small UAV with a mass of 150 kg is required to land on the ground with the help of a parachute with an impact velocity of 0.6 m/sec. What parachute projected area is needed for landing on sea level altitude?

5. A UAV with a mass of 180 kg is required to land on the ground with the help of a parachute. The parachute projected area is 6 m^2. Determine the impact velocity for landing on sea level altitude.

6. A very small UAV has a mass of 1 kg. A human operator with a maximum force of 80 N is arranged to launch this UAV. If his hand extends 75 cm, determine the release speed.

7. A human operator with a maximum force of 100 N is arranged to launch a very small UAV. His hand extends 70 cm, and the UAV release speed is required to be 6 m/sec. Determine the maximum possible mass of the UAV that can be launched.

CHAPTER 10

Payloads Selection/Design

10.1 INTRODUCTION

A UAV is designed to perform a given mission by carrying a particular payload. Thus, the payload is the object/part/equipment which the basic UAV provides a platform and transportation. The payload capacity of a UAV is a measure of size, weight, and energy available to perform the mission beyond the regular flight.

Payload similar to any object has weigh and volume. The UAV should have capacity (in space, and weight) to carry a payload. As the range and endurance of a UAV is increased, it can carry less payload. Table 10.1 summarizes characteristics of several large and small fixed-wing and rotorcraft UAS along with their payloads weight ratios. It is noticed that the payload weigh ratio (W_{PL}/W_{TO}) is from 8% (Killer Bee) to 28% (Black Eagle 50). The payload weight along with its volume are two important factors in payload selection process.

Table 10.1: Payloads weight ratio of several UAVs						
No	Name/ Designation	Endurance	Range	TO weight	Payload weight	W_{PL}/W_{TO}
1	RQ-4 Global Hawk	32 hr	22,800 km	32,250 lb	3,000 lb	0.093
2	Raven RQ-11B	90 min	10 km	4.2 lb	6.5 oz	0.097
3	Scan Eagle	23 hr	100 km	48.5 lb	13 lb	0.27
4	RQ-7B Shadow	7 hr	110 km	375 lb	100 lb	0.266
5	MQ-1C Gray Eagle	30 hr	3,750 km	3,600 lb	800 lb	0.22
6	A-160 Hummingbird	24 hr	4,023 km	6,500 lb	1,000 lb	0.15
7	MQ-8 Fire Scout	8 hr	203 km	3,150 lb	500 lb	0.16
8	Black Eagle 50	4 hr	260 km	77 lb	22 lb	0.28
9	Silver Fox	12 hr	32 km	28 lb	4 lb	0.14
10	Killer Bee	18 hr	100 km	250 lb	20 lb	0.08

Payloads are considered in two basic types—dispensable and non-dispensable. The non-dispensable payloads will remain with the UAV until the end of flight (e.g., sensors, and camera). The dispensable payloads will be dropped/delivered/launched during flight operation (e.g., armament for the military, and crop-spray fluid or firefighting materials for civilian use). There are two groups of sensors—sensor as a payload, and sensors necessary for navigation/guidance systems.

In this chapter, only payload sensors are addressed. The sensors which are necessary for a regular flight (such as the ones used in navigation/guidance systems) have been introduced in Chapters 5 and 6.

In this chapter, the definition of payload, types of payloads, and some UAV design considerations regarding the inclusion of payload is presented. In addition, the features of some popular payloads are described.

10.2   PAYLOAD DEFINITION

For a manned aircraft, payload is usually passenger, cargo, and store. The term initially arose from the load carried by an aircraft the transport of which was paid for by a customer. In contrast, for a UAV, the payload is defined as the sensor/equipment installed by the customer that performs a specific task. Basically, the aircraft (manned or unmanned) should be able to fly with the payload removed. This definition, of course, should avoid any misconception. Any equipment/sensor which is necessary for a regular flight will not be assumed as a payload. For instance, communication system, the autopilot, and navigation system sensors such as accelerometers, gyros, and magnetometers are not assumed as payload. Moreover, the elements/parts which are necessary for launch and recovery are not assumed as payload. For instance, in a pizza delivery UAV, only the pizza is the payload; even, the equipment to hold and release the pizza are not part of payload.

In some references, there is a broader definition for payload, which seems ambiguous. In those references, the payload covers all equipment and sensors which are carried by the structure, including avionics. Such definition makes it hard to compare two UAVS in terms of their useful payloads. In this book, any equipment and store which are necessary for flight, navigation, and recovery (flight-essential items) are not considered as payload.

There are mainly four types of payloads for a UAV: (1) cargo/freight, (2) reconnaissance, surveillance, monitoring (civil/military), (3) military payload (weapon), and (4) electronic warfare. In each type, there are a group of payloads which can be carried by a UAV for various objectives. The technical characteristics of some common/popular payloads will be presented in the next sections.

10.3   CARGO OR FREIGHT PAYLOAD

Cargo or freight are civil goods or product that is conveyed for commercial gains. Cargo may be stored in containers, or packed as a single item. The primary goal of carrying such payloads is mainly for delivery to a customer. The list of civilian payload for delivery by a UAV is countless; it includes items such as mails, books and pizzas, and packages and groceries. Due to safety regulations and cost of delivery, these payloads are not still common in the market.

There is broad interest in the tech world in using drones to make delivery fast and efficient. In the past few years, a number of companies tried to employ UAVs for the delivery of their products. In early 2016 [40], Google delivered Chipotle burritos at Virginia Tech. It has also negotiated with Starbucks about delivering with drones. And Domino's has delivered pizza in New Zealand via a drone.

In December of 2016, Amazon [40] kicked off a private trial of its highly anticipated UAV delivery program. Amazon said it would expand to dozens of customers near its British facility in the coming months and later grow to hundreds. The trial is limited to daytime deliveries when weather is suitable. Amazon is planning for making the deliveries in 30 min at no extra cost. A week earlier in Cambridge, Amazon used one of its autonomous drones for its first delivery, an Amazon Fire TV and a bag of popcorn. Amazon said it took 13 min from the customer clicking "order," to the package being delivered.

Moreover, in March of 2017, Amazon completed its first public demonstration of a Prime Air drone delivery in the U.S, ferrying sunscreen to attendees at an Amazon-hosted conference in California. It marks the first time one of the online retailer's autonomous UAV was flown for the public in the U.S. outside of Amazon's private property. In this mission, a quadcopter glides through the air carrying a box filled with sunscreen—weighing about four pounds—under its center body. The drone then touches down on a small landing pad in a field, where it releases the payload before vertically taking off. The mission was completed fully autonomously with Amazon's own software.

A widespread application of drone technology to deliver packages to customers' doorsteps requires a thorough analysis of the costs and benefits of this new pick-up and delivery system. Kharchenko and Prusov [41] reviewed the basic directions of unmanned aircraft systems applications in the civil field. The concepts of creation and organization of civil unmanned aircraft systems have been discussed. Moreover, the wide range of issues concerning the use of existing capacity for the design, manufacture and operation, with subsequent integration into the common air space have been addressed.

10.4   RECONNAISSANCE/SURVEILLANCE PAYLOAD

UAVs can carry various sensor packages, including electro-optical and infrared sensors, SAR, SIGINT, and multi- and hyperspectral imagers for reconnaissance/surveillance missions. There are basically three types of surveillance missions: (1) convoy over-watch, (2) route surveillance, and (3) fixed surveillance. One potential for applying UAVs to surveillance missions is to utilize a UAV for convoy over-watch as the convoy proceeds along its route. In addition, UAVs have an effective role in fixed-site security as an integral part of larger security and surveillance systems typically dominated by radars and fixed cameras. In military applications, a UAV may be flown in support of theater reconnaissance prior to the deployment of military forces. Moreover, UAVs are valuable for locating cargo that misses a drop zone. The river navigability using light detection and ranging (LIDAR) aboard a UAV is technically infeasible.

The reconnaissance and surveillance tasks put a premium on "real-time" sensors, or those whose results can be exploited immediately without the need for lengthy and time-consuming post-processing and annotation in order to render them intelligible. The requirement for real-time exploitation means that the options would be limited to electro-optical/infrared sensors, some varieties of Synthetic aperture radar (SAR), and Signals Intelligence (SIGINT). Other imaging systems exploiting different spectra cannot satisfy the real-time criterion.

Three primary functions of surveillance sensors are: (1) detection, (2) recognition, and (3) identification. Detection is defined as the activity to determining if there is an object of interest at a particular point in the target area. Recognition is the sensor determines if the object belong to a general class such as tree, bird, human, can, or tank. Identification is defined as determining the specific identity of an object, such as enemy soldier vs. own soldier, sedan car vs. van, and M1 Abram tank vs. T-92 tank. The success for target detection, recognition, and identification based on an image not only is function of the sensor sensitivity and resolution, but also due to the target signature. Target detection, tracking, recognition, and identification based on imaging are complicated processes and are affected by many parameters.

One of the most significant factors is image resolution. Imaging sensors produce images of varying degrees of resolution. These degrees of resolution are characterized by the NIIRS[15] with a numerical score, typically ranges from 1–9, where 1 represents very-low-resolution images and 9 very-high-resolution images. For instance, at an image with a score of 5, there is enough resolution at normal size to distinguish cars and buildings, and to identify planted crops. The available real-time sensors typically generate imagery with NIIRS ratings between 5 and 6.5. With the progress in miniaturization, we will probably see improvements in sensor technologies.

Two most common kinds of surveillance payloads (i.e., sensors) are daytime or night-vision cameras and radar. The features of these two surveillance sensors are described in this section.

10.4.1 CAMERA

A camera is an optical instrument (i.e., imaging sensor) for recording or capturing images, which may be stored locally, and/or transmitted to another location. The images may be individual still photos or sequences of images constituting videos/movies. The camera is a remote device as it senses a subject without physical contact. Factors affecting the quality of an image/video are resolution, focus, aperture, shutter speed, white balance, metering, film speed, and autofocus point. The area of the receptor of the camera and the number of picture elements, pixels, that it contains will determine the resolution of the image.

In general, there are three groups of cameras ranging from electro-optic (EO), Infra-Red (IR), and Laser Designator (LD). An electro-optic camera presents its output in a form that can be

[15] National Imagery Interpretability Rating Scale

interpreted by an operator as an image of what the sensor is viewing. Electro-optic cameras range from simple monochrome or color single-frame cameras through color TV cameras, low-light-level television, and thermal imaging video cameras to multi-spectral cameras. Optical, or "visible light" cameras operate in the 0.4–0.7 μm wave length range. For an IR camera, the image presented represents variations in the temperature and emissivity of the object in the scene. The infrared, or heat radiation, spectrum covers longer wavelengths—in the range of about 0.7–1000 μm.

In contrast, a radar sensor provides synthetic images, often includes false colors that convey information about target motion, and polarization of the returned signal. It is interesting to note that the Global Hawk (Block 20) is equipped with a synthetic aperture radar, an electro-optical camera and an infrared camera.

Some available camera classes in the market are: (1) rangefinder camera, (2) instant picture camera, (3) single-lens reflex, (4) twin-lens reflex, (5) large-format camera, (6) medium-format camera, (7) subminiature camera, (8) movie camera, camcorders, (9) digital camera, (10) panoramic camera, and (11) VR camera.

It is worth comparing two cameras; a simple and cheap camera for RC modelers, and an advanced one for the Mars Rover. The Wi-Fi Enabled FPV High-Definition Ominus Camera with dimensions of 73 × 28 × 14 mm, and a range of 100 mm has a Resolution: 1280 x 720 DPI. The price tag of this camera which takes both photos and videos is about $63.

There are nine cameras hard-mounted to the rover: two pairs of black-and-white Hazard Avoidance Cameras in the front, another two pair mounted to the rear of the rover, and the color Mars Descent Imager. There is one camera on the end of a robotic arm that is stowed in this graphic; it is called the Mars Hand Lens Imager (MAHLI). The MAHLI is similar to that of consumer digital cameras, with an autofocus ability. Some features are as follows: image size: up to 1600 × 1200 pixels; image resolution: possibility of 13.9 microns/pixel; focal length: in focus from 18.3 mm at the closest working distance to 21.3 mm at infinity; focal ratio: from f/9.8 to f/8.5; field of view: from 34° to 39.4°; memory: 8 gigabyte flash memory storage; 128 megabyte synchronous dynamic random access memory (SDRAM); and HD video: 720p. The cost of these advanced cameras top millions of dollars.

A long-range HALE UAV for a typical reconnaissance operation to detect a target will need a wide FOV lens to focus onto a receptor with a large number of pixels. This type of camera is usually large, heavy, more complex, and requiring great electrical power for its operation. However, a close-range UAV will need a lighter, low-cost, smaller, and less capable camera.

To obtain a 360° azimuth field of view, a rotatable turret should be mounted beneath the UAV. In addition, the elevation and roll gimbals, carrying the sensor(s), should be mounted with their actuators within the turret. Figure 10.1 illustrates five CONTROP's medium range stabilized EO/IR cameras. These cameras are 3 axis gimbal gyro-stabilized payloads for medium range (up to 3–5 km).

Figure 10.1: CONTROP's medium range stabilized EO/IR Cameras (EO/IR Gyro-stabilized camera payloads by CONTROP Precision Technologies Ltd.).

10.4.2 RADAR

Radar (RAdio Detection And Ranging) is an object-detection device that uses electromagnetic (e.g., radio) waves to measure the range (distance), angle, or velocity of an object. Four main elements of a radar are the: (1) transmitter, (2) receiver, (3) signal processor, and (4) antenna. Often, the transmitting antenna and the receiving antenna are the same. A similar device to radar to make use of other parts of the electromagnetic spectrum is LIDAR, which uses ultraviolet, visible, or near infrared light from lasers rather than radio waves.

A radar provides its own source of energy, hence it does not depend on reflected light or heat emitted from the target. The wave which is sent by radar could be either continuous or pulsed. A radar transmitter emits radio signals in predetermined directions. If the transmitter turns, the signal direction is turned as well. If the Earth were a perfectly flat horizontal plane, the signal would come only from the closest point, and would be a true measure of altitude. However, the Earth is not smooth, and energy is scattered back to the radar from all parts of the surface illuminated by the transmitter.

A transmitter generates the radio signal with an oscillator such as a magnetron. A receiver is an electronic device that receives radio waves and converts the information carried by them to a usable form. An antenna is an electrical device which converts electric power into radio waves and vice versa. An antenna either receives energy from an electromagnetic field or transmits (i.e., radiates) electromagnetic waves produced by a high frequency generator, or returned from a target. Antenna is a structure which serves as a transition between wave propagating in free space, and the fluctuating voltages in the circuit to which it is connected (i.e., transmitter). The most frequently used type of an antenna is in the form parabolic dish (for example, see Figure 9.1).

The radar signals that are reflected back are employed by the receiver and will be sent to the radar processor. The radar measures distance by determining the time required for a radio wave to travel to and from a target. The radar frequencies for UAVs are often from 9–35 GHz. The electromagnetic waves are travelling with speed of light. Thus, when a target is located at a distance of even 100 km away from the radar, the measurement is assumed real-time. The performance of most radar is limited by the ability of radar to separate targets from clutter. If the radar beam is way larger the target dimensions, the performance can be enhanced.

A particular type of radar is the radar altimeter which is often used in aircraft during bad-weather landings. Radar altimeter is much more accurate, and more expensive than the pressure altimeters. They are an essential part of many blind-landing and navigation systems and are used over mountains to indicate terrain clearance. Special types are used in surveying for quick determination of profiles. Radar altimeters have been is use on various spacecraft, to measure the shape of the geoid and heights of waves and tides over the oceans.

A radar operates based on the concept of Doppler effect (or the Doppler shift) which is the change in frequency or wavelength of a wave signal for an observer moving relative to its source. The relative motion that affects the observed frequency is only the motion in the LOS between the source and the receiver. When the speeds of source and the receiver relative to the medium (i.e., air) are lower than the velocity of waves in the medium, the relationship between observed frequency (f) and emitted frequency (f_o) is given by:

$$f = \frac{c + v_r}{c + v_s} f_o \qquad (10.1)$$

where c is the wave velocity (about 300,000 km/sec), v_r is the velocity of the receiver (i.e., UAV), v_s is the velocity of the source (i.e., target). When the transmitted and received frequencies are measured, the target velocity is calculated by the radar processor using this equation. The frequency of a sinusoidal signal is expressed in terms of the wavelength (λ):

$$f = \frac{c}{\lambda}$$

$$(10.2)$$

Radar equation relates the returning power, P_r and transmitted power, P_t:

$$P_r = \frac{P_t\, G_t\, A_r\, \sigma F^4}{(4\pi)^2\, R^4} \qquad (10.3)$$

where G_t denotes the gain of transmitting antenna, A_r the effective aperture, σ radar cross section, F the pattern propagation factor, and R the distance between UAV radar and the target. The equation indicates that the received power declines as the fourth power of the range, which means that the received power from distant targets is relatively very small.

A type of modern radar to provide finer spatial resolution than conventional beam-scanning radars is Synthetic Aperture Radar (SAR). The SAR, is a side-looking radar system that electronically simulates an extremely large antenna or aperture, and generates high-resolution remote sensing imagery. In this process, known as coherent detection; the phase of a return signal is compared with that of the transmitted signal. The details of radar design are beyond the scope of this book; for details see references such as Richards et al. [43].

10.5   SCIENTIFIC PAYLOADS

There is a long list for non-reconnaissance/surveillance payloads which may be utilized by UAVs for various scientific missions. Some significant ones are spectrometers, radiation detectors, environmental sensors, and atmospheric sensors. A brief introduction to these sensors are presented here.

- A **spectrometer** analyzes the spectrum of light to identify the chemical elements of the target. The technology may also be used in other extreme conditions such as nuclear reactors and the sea floor. For instance, Mars Rover is equipped with a rock-zapping laser and telescope. It produces more than a million watts of power focused for five one-billionths of a second to hit targets up to seven meters away. The result is a glowing ionized gas which it observes with the telescope. The light, then, is analyzed by spectrometer.

- A **radiation detector** monitors high-energy atomic and sub atomic particles coming from the sun, from distant supernovae and other sources to determine whether conditions are favorable to life and preserving evidence of life. Naturally occurring radiation is harmful to microbes and human life. The Earth atmosphere has a protective magnetic field against radiation.

- An **environmental sensor** records environmental information such as carbon monoxide, water vapor, aerosols, ozone, gases composing air, air viscosity, chemicals in the air, and ultraviolet radiation. The wind speed and direction, temperatures, and humidity are measured by electronic devices. The atmospheric pressure is measured by a barometer. Some other atmospheric phenomena include rain, icing, snow, wind, gust, turbulence, hurricane, tornado, thunderstorm, and lightning.

- An **atmospheric sensor** records atmospheric information such as wind speed, wind direction, air pressure, relative humidity, air temperature, ground temperature. The wind speed and direction, temperatures, and humidity are measured by electronic devices. The atmospheric pressure is measured by a barometer.

The scientific payloads are manufactured in various configurations, sizes, and weights. For details characteristics, refer to the manufacturers websites or design your own.

10.6 MILITARY PAYLOAD (WEAPON)

In general, there are two groups of military missions—delivery of lethal warhead to a target and military operations with no lethal warhead. The first group involves and operations such as strike, interception, and attack (e.g., improvised explosive device; and landmine detection and destruction). The latter includes missions such as military reconnaissance and surveillance, electronic intelligence, and target designation by laser. Please note that the guided weapons (e.g., cruise missiles) are not covered in this book. Payloads relating to the latter group have been discussed in Section 10.4. In this session, only weapon payloads are addressed. The UAV which are utilized for such payloads are in the category of unmanned combat air vehicle (UCAV).

The weapon payloads are primarily include stores and munitions such as missiles, rockets, and bombs. These weapons are frequently attached underneath the UAV, and the time of need, they are released or dropped, or launched. The number of weapons required for a mission can be traded off against longer range, lighter weight, smaller size, and lower cost. If the radius of action is shorter than the designated range, more payload can be carried with less fuel.

An important requirement for an armed UAV is that it should be capable of taking off with a useful load of weapons. There are a number of factors affecting the capability of a UAV to carry a lethal payload including the payload weight. In general, the Army missiles are less heavy than the Air Force missiles. For this reason, the Predator UAV was originally equipped with Hellfire missiles; simply because the Air Force air-to-ground missiles are two heavy for the Predator to carry.

There are variety of weapon payloads that can be carried by a combat UAV in terms of weight, size, configuration, type of warhead, and the mission. In this section, as an example only one missile is introduced. For more information about any other aerial weapons, you may consult with DOD publications.

The AGM-114 Hellfire (Figure 10.2) is an air-to-surface missile initially developed for anti-armor application, but later models were developed for precision strikes against other target types. The basic Hellfire has a length of 163 cm, a mass of 45 kg, and is equipped with a solid fuel rocket motor, and a semi-active laser homing guidance system. It can target tank, armored vehicles, and even individuals. The range of this missile is 8 km. A Predator UAV can carry up to two Hellfire missiles; and have been used in a number of targeted killings of high-profile individuals.

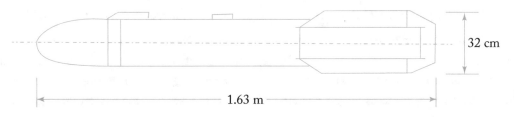

Figure 10.2: A military payload, AGM-114 Hellfire missile.

Air-to-surface missiles are sophisticated and very expensive weapon payloads; and currently controlled (i.e., launched) from a ground control station. The design of a UAV to carry such payload is a very challenging task. There are mechanical and electrical interfaces, extreme care must be taken to guarantee the safety of UAV over the launch period.

10.7 PAYLOAD INSTALLATION

When the payloads for a UAV is selected, the installation is the next step which may even affect the UAV configuration. The payload will require at least a space and power allocation. The payload installation has various aspects; four of which are briefly addressed in this section: (1) payload location, (2) payload aerodynamics (in mounted externally), (3) payload-structure integration, and (4) payload stabilization.

10.7.1 PAYLOAD LOCATION

Generally, there are two locations for a payload—inside UAV (e.g., podded) and outside on UAV. If a payload is considered to be installed inside a UAV (e.g., a radar), the space management and center of gravity (cg) adjustment are required. In contrast, if the payload is installed out of the UAV (e.g., below fuselage), lofting technique is employed to reduce payload drag. When a payload is installed inside the body, the designer must make sure that the payload performance is not negatively affected. For instance, the design of radome (i.e., its material, thickness, and curvature) will influence the performance of a radar.

Two important factors affecting the payload location are the payload mass and dimensions, which is subsequently driven by the needs of the operational tasks. The type and performance of the payloads can range from:

1. relatively simple sub-systems consisting of an un-stabilized video camera with a fixed lens having a mass as little as 200 g, to

2. a video system with a greater range capability, employing a longer focal length lens with zoom facility, gyro-stabilized and with pan and tilt function with a mass of probably 3–4 kg, to

3. a high-power radar having a mass, with its power supplies, of possibly up to 1,000 kg.

The distribution of aircraft weight will influence the airworthiness and performance via two aircraft parameters—aircraft center of gravity (cg) and aircraft mass moment of inertia. Aircraft must be stable, controllable and safe for all allowable aircraft cg locations during flight envelope. One of the primary concerns during aircraft design process related to the payload location; is the UAV weight distribution. The distribution of UAV weight (sometimes it is referred to as weight and balance) will greatly influence airworthiness as well as aircraft performance. The aircraft center of gravity is the cornerstone for aircraft stability, controllability, and trim analysis.

If the payload is expendable (e.g., weapon), the location of payload will influence the cg range too. When the military payload is delivered, the UAV cg will move to a new location. As a rule of thumb, the best aircraft cg location is to be around wing-fuselage aerodynamic center. The distance between the most forward and most aft center of gravity limits is called the center of gravity range or limit along x axis. In general the location of a payload must be such that the entire UAV cg is within the acceptable range (both initially, and after the payload is released). For technique to distribute weight, and to determine the UAV range of center of gravity, refer to Chapter 11 of Sadraey [37].

10.7.2 PAYLOAD AERODYNAMICS

If the payload is installed out of the UAV (e.g., below fuselage), techniques (e.g., lofting) should be employed to reduce payload drag. This is the case for almost all weapon payloads and cameras. In designing a cover for payload, both aerodynamics and overall weight need to be considered. A small cover is light, but may create a high drag. In contrast, a large cover may decrease drag, but it will have a higher weight. Hence, a trade-off is necessary. Techniques to determine UAV drag are presented by references such as Sadraey [38].

Keep the aerodynamic drag of the UAV as low as possible commensurate with the practical installation and operation of the payload, and radio antennae. Where the payload is connected to the fuselage, some forms of filleting (e.g., fillet) are required to avoid flow separation and turbulence. The exact shape may be determined by wind tunnel experiments. Figure 10.1 demonstrates the cover of a few CONTROP's Cameras. For fundamentals of aerodynamics, please refer to references such as Anderson [44], and Shevell [45].

10.7.3 PAYLOAD-STRUCTURE INTEGRATION

Whether the payload in inside or outside of the UAV, the structural integrity of the UAV must be maintained such as the various flight loads (e.g., gust, aerodynamic, and weight) and stresses (e.g., normal, shear, bending) are handled safely by the structure. The load is significantly in high-g turn and hard landing. When the payload is installed inside the UAV, a special mounting should be designed to allow the payload to perform properly.

If the payload is installed out of the UAV (e.g., below fuselage), provisions are required for mounting payloads on so-called hard-points" under wing or fuselage. The structure must be designed to provide mounting points capable of supporting the launch rails or bomb racks. Installation of a payload outside of structure requires a cutout by reinforcing the neighboring structural elements (e.g., stiffener, frame). When a missile is launched, the reaction force is passed from rails to the structure. Thus, the structural elements near the rails must be reinforced. The UAV structural design is out of scope of this book, you may refer to references such as Megson [47] for more details.

In stealth UAV, a concern for the installation of a payload is the radio/radar signature. There are basically three methods of minimizing the reflection of pulses back to a receptor:

1. to manufacture appropriate areas of the UAV from radar-translucent material such as Kevlar or glass composite as used in radomes which house radar scanners;

2. to cover the external surfaces of the aircraft with RAM (radar absorptive material); and

3. to shape the aircraft externally to reflect radar pulses in a direction away from the transmitter.

10.7.4 PAYLOAD STABILIZATION

Some payloads such as cameras need a stable location, in order to have an acceptable performance. Moreover, in order to obtain a quality image, the sight-line must be directed to angular space at whichever angle is demanded. These two requirements are achieved by a stabilization system, which is a closed-loop system which the motion of the base is sensed and fed back. A controller sends commands to actuators to rotate the gimbals. So, any motion of the base is nullified by the gimbals, hence the camera remains still. For instance, if the base is pitched up 5°, a gimbal is pitched down by -5°.

A camera may be nose-mounted or in turret-mounted configuration. Stabilization of the sight-line is necessary to off-set the effects of both UAV movements and displacements of the aircraft due to gusty/turbulent air. To make the lens field of view narrower, the sight-line must be held more precisely. The three-axes (x, y, and z) stability is the highest possible way to provide the sight-line stability.

There are two basic techniques of maintaining a stabilized sight-line—the camera assembly is mounted on a stabilized platform with attitude and rate sensors, or the camera assembly is mounted on gimbals remote from the attitude sensors. In the first technique, the camera with its gimbals is mounted on a platform which is gyro-stabilized. Fast-reacting actuators are ready to correct any movement away from the horizontal. Thus, the sight-line keeps its direction of view relative to the horizontal base. In the second technique, the whole camera is contained within an exchangeable payload module with the gimbals' base fixed to the structure of that module to provide a stable picture. In general, the second method offers a better maintainability, while it is less expensive. The aircraft attitude data is coming from the navigation system and used to drive the gimbal actuators.

10.8   PAYLOAD CONTROL AND MANAGEMENT

Weapon payloads have mechanical and electrical interfaces, extreme care must be taken to guarantee the safety of the UAV at the time of launch. An external payload (e.g., missile, camera) has an electrical connection to the UAV via a connector; referred to as umbilical. The connector can be plugged into the payload and will come free when the payload is released/launched. The configuration of the connector assembly must be such that it provides a safe separation for both payload and UAV. Various data is exchanged through an umbilical. A few examples are imagery from camera/seeker, control signals, and launch signals.

In addition to maintaining control and stability of the aircraft, it is just as important to achieve that for the payload. Control of the UAV is needed to get the vehicle over the target area, but will be useless, unless the payload is properly controlled. The payload control is achieved using a system which is either part of the aircraft flight control system or by using a separate module. For a camera, the output of the control process, is to bringing the sight-line accurately on to the target and keeping it there. The gyro-stabilization of the sight-line is a function of the camera control system.

The integration of the payload control and stability systems and flighty control system is a challenging task. In an integrated system, the same set of gyros will support the control and stabilization of both aircraft and payload. The integrated system manages two sets of coordinate axes, those of the aircraft and those of the payload, even though the latter is fixed within the aircraft. Thus the payload sight-line is at the target while the UAV is flying in another direction.

Another aspect of the payload control is the data management, processing, and transmission. Images and/or data from the various payloads must be transmitted to the ground control station. The output from sensors within the payload will require processing and converting to radio signals. Different sensors will require different amounts of radio bandwidth and at different data rates. Thus, it is necessary that as much processing of the signals as is practical is carried out within the UAV in order to reduce signal loads on the communication system. In general, the basic fundamentals of payload control is very similar to flight control system. Both have feedback, controller, and measurement devices. The theory behind control is covered in Chapter 4, so they are nor repeated here.

10.9   PAYLOAD SELECTION/DESIGN CONSIDERATIONS

Selection of the payload has a significant impact on the UAV configuration; since it is tied with UAV mission and performance. A UAV designer must be aware of the ambiguity about payload capacity, and use a consistent definition. Sometimes, the payload function in-flight may interfere with the UAV flight management. This interference should be accurately planned and managed in order not to compromise the flight safety.

However, certain payloads may also require access to UAV system data, such as position, attitude, or airspeed. Thus, a mechanism must exist to provide UAV data to the payload in a manner such that the failure of the payload cannot impact the safety of the UAV's own systems. In some cases, a UAV is required to be capable of a range of different roles, and carrying various payloads and different capability in the mission. Thus, a compromises will be necessary in designing payload compartment.

In the preceding sections, features of various payloads and techniques to control, handle, and manage have been described. Moreover, the installation issues including drag reduction, payload-structure integration, and optimum payload location were addressed. In selection/design process of payload(s) for a UAV, one must select the type of payloads, select the payload devices, and then conduct some calculations and performance/stability analysis. In general, the primary factors affecting the selection/design process of payload are as follows: (1) manufacturing technology, (2) accuracy and precision, (3) volume, (4) environmental factors, (5) reliability, (6) life-cycle cost, (7) UAV configuration, (8) stealth requirements, (9) maintainability, (10) energy consumption, (11) integration with structure system, (12) aerodynamic considerations, (13) processor, (14) interference with flight safety, (15) compatibility with control system, (16) weight, and (17) radio/radar signature. Figure 10.3 illustrates a flowchart that represents the payload(s) selection/design process.

In general, the design process begins with a trade-off study to establish a clear line between cost and performance (i.e., accuracy) requirements; and ends with optimization. The designer must decide about two items—select type of payload and select the payload devices. Most payload devices may be ordered/purchased from off-the-shelf commercial market. After conducting the calculation process, it must be checked to make sure that the design requirements are met. A very crucial part of the design process is to integrate the payloads with structure and control system (i.e., a consistent autopilot). If the complete payloads (e.g., camera) is selected/purchased, the integration process must be still conducted. This includes activities such as structural installation, management, interfaces, and electric power requirements. More details is outside the scope of this work.

10.10   QUESTIONS

1. Define payload.

2. Provide at least five types of payloads that a UAV may carry for various missions.

3. What factors are affecting the quality of an image/video in a camera?

4. What provisions are necessary in designing payload compartment?

5. What are the primary factors affecting the payload selection/design process?

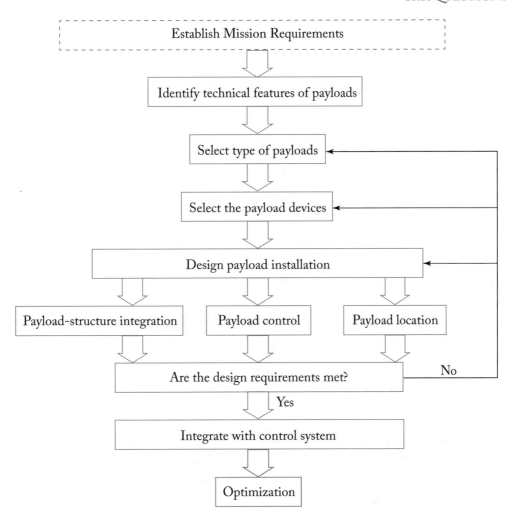

Figure 10.3: Payload selection/design process

6. Name four scientific payloads.

7. Define the Doppler effect.

8. What are the typical radar frequencies used for UAVs?

9. What can be concluded from radar equation?

10. What is the typical outcome of the Doppler effect?

11. What factors influence the resolution of a camera?

12. Describe the synthetic aperture radar.

13. What is the speed of a radio wave?

14. Describe receiver, transmitter, and antenna in brief.

15. What is LIDAR?

16. Provide three examples for the weapon payloads.

17. Name four aspects of the payload installation.

18. What is the functions of an payload umbilical?

19. Describe the overall payload selection/design process.

10.11   PROBLEMS

1. A sound wave is moving with the frequency of 800 kHz. Assume the sound travels at 290 m/sec, determine the wavelength.

2. An antenna of a UAV transmitter radiates a radio signal with the frequency of 20 GHz. Assume the wave travels with a speed of 300,000 km/sec, determine the wavelength.

3. A UAV flying at an altitude of 2 km is directly behind and closing at a horizontal range of 10 km from a carrier. The UAV with a speed is 100 m/sec parallel to the ground, is tracking the ship with a radar with a frequency 12 GHz. The ship cruises at 5 m/sec. If the speed of light is 3×10^8 m/sec, calculate:

   a. the radar frequency detected by the aircraft carrier,

   b. the echo frequency detected by the UAV,

   c. the Doppler shift between the echo and source frequency,

   d. wave length.

4. A UAV flying at an altitude of 3 km is directly behind and closing at a horizontal range of 10 km from a carrier. The UAV with a speed is 50 m/sec parallel to the ground, is tracking the ship with a radar with a frequency 20 GHz. The ship cruises at 7 m/sec. If the speed of light is 3×10^8 m/sec, calculate:

   a. the radar frequency detected by the aircraft carrier,

b. the echo frequency detected by the UAV,

c. the Doppler shift between the echo and source frequency,

d. wave length.

5. A fixed sound source emits a sound with a frequency of 1 MHz on a day when the speed of sound in air is 340 m/sec, and there is no wind. What is the frequency you will receive if you move toward the source at 20 m/sec?

6. A moving sound source emits a sound with a frequency of 2 MHz on a day when the speed of sound in air is 330 m/sec, and there is no wind. What is the frequency you will receive if the source moves toward you at 30 m/sec? Assume you are stationary.

Bibliography

1. Roskam, J. *Airplane Flight Dynamics and Automatic Flight Controls*, 1997, DARCO. 54, 71

2. Stevens, B. L., Lewis, F. L., and Johnson, E. N. *Aircraft Control and Simulation*, 3rd ed., John Wiley, 2016. 53, 94

3. Federal Aviation Regulations, Part 23, Airworthiness Standards: Normal, Utility, Aerobatic, and Commuter Category Airplanes, Federal Aviation Administration, Department of Transportation, Washington DC. 63

4. Federal Aviation Regulations, Part 25, Airworthiness Standards: Transport Category Airplanes, Federal Aviation Administration, Department of Transportation, Washington DC. 63

5. MIL-STD-1797, Flying Qualities of Piloted Aircraft, Department of Defense, Washington DC, 1997. 63

6. MIL-F-8785C, Military Specification: Flying Qualities of Piloted Airplanes, Department of Defense, Washington DC, 1980. 72

7. Mclean, D. *Automatic Flight Control Systems*, Prentice-Hall, 1990.

8. Nelson, R. *Flight Stability and Automatic Control*, McGraw Hill, 1989.

9. Hoak, D. E., Ellison D. E., et al, "USAF Stability and Control DATCOM," Flight Control Division, Air Force Flight Dynamics Laboratory, Wright-Patterson AFB, Ohio, 1978.

10. Jackson, P., Jane's All the World's Aircraft, Jane's information group, Various years.

11. Dorf, R. C., Bishop R. H., *Modern Control Systems*, 13th ed., Pearson, 2017. 64, 65, 73

12. Ogata, K., *Modern Control Engineering*, 5th edition, Prentice Hall, 2010. 64, 65, 73

13. Doyle, J. C. and Glover, K., State-space solution to standard H2 and H∞ control problems, *IEEE Transactions on Automatic Control*, Vol. 34, No. 8, August 1989. 74

14. Anderson, B. D. O. and J. B. Moore, *Optimal Control: Linear Quadratic Methods*, Dover, 2007. 51, 73

15. Phillips, C. L, Nagle T., and A. Chakrabortty, *Digital Control System Analysis & Design*, 4th ed., Pearson, 2014. 51

16. Blakelock, J. H., A*utomatic Control of Aircraft and Missiles*, 2nd ed., John Wiley, 1991. 112

17. http://manuals.hobbico.com/snn/snna1051-manual.pdf.

18. Pastrick, H.L., S.M. Seltzer, and M.E. Warren. Guidance laws for short-range tactical missiles, *Journal of Guidance, Control, and Dynamics*, Vol. 4, No. 2 (1981), pp. 98-108. DOI: 10.2514/3.56060. 107

19. Siouris, G. M., *Missile Guidance and Control Systems*, Springer, 2004.

20. Manley Butler, C., Jr. and Troy Loney, Design, Development and Testing of a Recovery System for the Predator UAV, *13th AIAA Aerodynamic Decelerator Systems Technology Conference*, May 15-19, 1995, Clearwater Beach, FL.

21. Ben-Asher, J. Z. and Yaesh, I., *Advances in Missile Guidance Theory*, American Institute of Aeronautics and Astronautics, Reston, VA, 1998. 119

22. Zarchan, P., *Tactical and Strategic Missile Guidance*, 6th ed., American Institute of Aeronautics and Astronautics, Reston, VA, 2013. 119

23. Palumbo, N., F., Blauwkamp, R. A., and Lloyd, J. M. Basic Principles of Homing Guidance, *Johns Hopkins APL Technical Digest*, Vol. 29, No. 1, 2010. 119

24. Grewal M. S., Andrews, A. P., and Bartone C. G., *Global Navigation Satellite Systems, Inertial Navigation, and Integration*, 3rd ed., Wiley, 2013. 95

25. Toolev M. and Wyatt D., *Aircraft Communications and Navigation Systems: Principles, Maintenance and Operation*, Routledge, 2007.

26. Ahmed El-Rabbany, *Introduction to GPS: The Global Positioning System*, 2nd ed., by 2006, Artech House Publishers.

27. Federal Aviation Regulations, Part 107, Operation and Certification of Small Unmanned Aircraft Systems, Federal Aviation Administration, Department of Transportation, Washington DC, 2016.

28. Tso, K. S., Tharp, G. K., Tai, A. T., Draper, M. H., Calhoun, G. L., and Ruff, H. A., Human Factors Testbed for Command and Control of Unmanned Air Vehicles, *22nd Digital Avionics Systems Conference*, Indianapolis, IN, October 2003. 144

29. Salyendy, G., *Handbook of Human Factors and Ergonomics*, 3rd ed., Wiley, 2006. 144

30. Berger, A. S., *Embedded Systems Design*, CMP Books, 2002. 125

31. Cady, F. M., *Microcontrollers and Microcomputers*, Oxford University Press, 1997. 124, 125

32. Ball, S., *Analog Interfacing to Embedded Microprocessors*, Newnes, 2001. 133

33. Miller, G. H., *Microcomputer Engineering*, Pearson Prentice Hall, 3rd ed., 2004. 126

34. Graham, D. L., *C Programming Language: A Step by Step Beginner's Guide to Learn C Programming in 7 Days*, 2016. 131, 134

35. Wilmshurst, T., *Designing Embedded Systems with PIC Microcontrollers, Principles and Applications*, 2nd ed., Newnes, 2009. 127

36. ArduPilot 2.x manual, ArduPilot Development Team, http://ardupilot.org, 2003. 134

37. Sadraey, M., *Aircraft Design; A Systems Engineering Approach*, Wiley, 2012. DOI: 10.1002/9781118352700. 27, 157, 181

38. Sadraey, M., *Aircraft Performance Analysis*, CRC Press, 2017. 159, 181

39. Frenzel, L., *Principles of Electronic Communication Systems*, 4th ed., McGraw-Hill. 153

40. CNN, Washington, December 14, 2016, http://money.cnn.com. 173

41. Kharchenko, V. and Prusov, D., Analysis of unmanned aircraft systems application in the civil field, in *Transport*, Taylor and Francis, 2012. DOI: 10.3846/16484142.2012.721395. 173

42. Peters, J. E., Seong, S., Bower, A., Dogo, H., Martin, A. L., and Pernin, C. G., *Unmanned Aircraft Systems for Logistics Applications*, RAND, 2012.

43. Richards, M. A., Scheer J. A., and Hilm W. A., *Principles of Modern Radar: Basic Principles*, SciTech Publishing, 2010. DOI: 10.1049/SBRA021E. 178

44. Anderson, J. D., *Fundamentals of Aerodynamics*, McGraw-Hill, 5th ed., 2010. 30, 181

45. Shevell, R. S., *Fundamentals of Flight*, Prentice Hall, 2nd ed., 1989. 30, 181

46. Jackson, P., *Jane's All the World's Aircraft*, Jane's information group, Various years. 4

47. Megson, T., *Aircraft Structures for Engineering Students*, 5th ed., 2012. 32, 182

48. Roskam, J., *Lessons Learned in Aircraft Design*, 2007, DAR Corporation.

49. Roskam, J., *Roskam's Airplane War Stories*, 2006, DAR Corporation.

50. Sadraey, M., *Robust Nonlinear Controller Design for Complete UAV Mission*, VDM Verlag Dr. Müller, 2009. 51

51. Anonymous, Literature Review on Detect, Sense, and Avoid Technology for Unmanned Aircraft Systems, DOT/FAA/AR-08/41, National Technical Information Service, 2009. 81

52. Dalamagkidis, K., Valavanis, K. P., and Piegl, L. A., *On Integrating Unmanned Aircraft Systems into the National Airspace System: Issues, Challenges, Operational Restrictions, Certification and Recommendations*, International Series on Intelligent Systems, Control, and Automation: Science and Engineering, Springer-Verlag, 2009, vol. 26. 81

53. Schwartz, C.E., Bryant, T.G., Cosgrove, J.H., Morse, G.B., and Noonan, J. K., A radar for unmanned air vehicles, *The Lincoln Laboratory Journal*, Vol. 3. No. 1, 1990. 119

54. *Unmanned Systems*, May/June 2004. 3

55. Butler, M. C. and Loney, T., Design, Development and Testing of a Recovery System for the Predator UAV, AIAA 95-1573, *13th AIAA Aerodynamic Decelerator Systems Technology Conference*, May 15-19, 1995, Clearwater Beach, FL. 167

Author Biography

Dr. Mohammad H. Sadraey is an associate professor in the College of Engineering at Southern New Hampshire University (SNHU), Manchester, New Hampshire. Dr. Sadraey's main research interests are in aircraft design techniques, aircraft performance, flight dynamics, and design and automatic control of unmanned aircraft. He earned his M.Sc. in aerospace engineering in 1995 from RMIT, Melbourne, Australia, and his Ph.D. in aerospace engineering from the University of Kansas, Kansas, in 2006. Dr. Sadraey is a senior member of the American Institute of Aeronautics and Astronautics (AIAA), Sigma Gamma Tau, and the American Society for Engineering Education (ASEE). He is also listed in *Who's Who in America*. He has more than 20 years of professional experience in academia and industry. Dr. Sadraey is the author of three other books, including *Aircraft Design: A Systems Engineering Approach* published by Wiley publications in 2012, and *Aircraft Performance* by CRC in 2016.

Printed in the United States
by Baker & Taylor Publisher Services